OUR ENVIRONMENTAL HANDPRINTS

Recover the Land, Reverse Global Warming, Reclaim the Future

JON R. BIEMER

ROWMAN & LITTLEFIELD
Lanham • Boulder • New York • London

Published by Rowman & Littlefield
An imprint of The Rowman & Littlefield Publishing Group, Inc.
4501 Forbes Boulevard, Suite 200, Lanham, Maryland 20706
www.rowman.com

86-90 Paul Street, London EC2A 4NE

British Library Cataloguing in Publication Information Available

Library of Congress Cataloging-in-Publication Data

Names: Biemer, Jon R., 1950– author.
Title: Our environmental handprints : recover the land, reverse global warming, reclaim the future / Jon R. Biemer.
Description: Lanham : Rowman & Littlefield, [2021] | Includes bibliographical references and index.
Identifiers: LCCN 2020049487 (print) | LCCN 2020049488 (ebook) | ISBN 9781538140659 (hardcover) | ISBN 9781538185483 (paper) | ISBN 9781538140666 (epub)
Subjects: LCSH: Environmental protection—Citizen participation. | Sustainable living. | Environmental responsibility.
Classification: LCC TD171.7 .B53 2021 (print) | LCC TD171.7 (ebook) | DDC 304.2—dc23
LC record available at https://lccn.loc.gov/2020049487
LC ebook record available at https://lccn.loc.gov/2020049488

To my wife, Willow Dixon,
a proactive partner in many of my adventures in sustainability.

To my father, Richard Biemer, who instilled in me the love of trees.

To those who created Handprints upon which we build.

To those who will follow through with Handprints of their own.

And there's a wonderful parable in the New Testament: The sower scatters seeds. Some seeds fall in the pathway and get stamped on, and they don't grow. Some fall on the rocks, and they don't grow. But some seeds fall on fallow ground, and they grow and multiply a thousandfold. Who knows where some good little thing that you've done may bring results years later that you never dreamed of?

—Pete Seeger*

* Amy Goodman, "'We Shall Overcome': Remembering Folk Icon, Activist Pete Seeger in His Own Words and Songs," https://www.democracynow.org/2014/1/28/we_shall_overcome _remembering_folk_icon.

CONTENTS

ACKNOWLEDGMENTS

Emé Alter, Natalia Blackburn, Kenneth Stotler, and Karl Vischer have walked this path with me for a very long time—thank you. Thank you as well to my agent, Lisa Hagan, and my book coach, Mark Malatesta, for their wise persistence.

The following people generously provided manuscript review: Christine Brautigam, Victoria Clevenger, Robin Cordova, Willow Dixon, Pamela Edwards, Doug Haines, Gayle Highpine, David Holladay, Sarah Horn, Diana Jones, Bob Leonard, Christine Lightcap, Milt Markewitz, Craig Moro, Linda Neale, Vera Sinnock, Karl Vischer, Lloyd Vivola, and Brian Wagner.

Also, thanks to Gregory Norris for giving me the opportunity to share my emerging Handprint ideas on the Handprinter website.

In this way, *all* of you own a part of the collective Handprint this book represents.

INTRODUCTION:
OUR CHILDREN ARE WATCHING

This world we are leaving to our children concerns you. You might have heard sixteen-year-old Greta Thunberg's speech to the United Nations: "My message [to world leaders] is that we'll be watching you. . . . Entire ecosystems are collapsing. We are in the beginning of a mass extinction, and all you can talk about is money and fairy tales of eternal economic growth. How dare you!" You take those words seriously, in the spirit of a worthy challenge. You already recycle and use reusable shopping bags. You turn down the thermostat at night. You shifted away from incandescent lighting a while back. Your car, if you have one, is more efficient than the last one. You may have a smaller Ecological Footprint than your neighbors.

Even so, it feels like you can only go so far in reducing your impact on the environment. Children need to be housed, fed, and schooled and relationships nurtured. The job has its constraints. You need to fly cross-country to visit family. Perhaps you feel plateaued in terms of what you can do for the environment.

This book answers the diminishing-return challenge we all experience when trying to reduce our resource consumption.

This book emphasizes a positive way of thinking—the Handprint. Simply put, the Handprint symbolizes the good we do. The potential benefit from our Handprints is unlimited. Handprints have a tendency to be self-perpetuating. Take planting a tree, for example. You carefully put a seedling into the ground, water it, and then leave it alone. From then on it pulls carbon dioxide from the air for as long as it lives. Entrepreneurs offering environmentally friendly products, like Patagonia clothing, have learned to profit from taking care of our planet (chapter 2).

I capitalize "Handprint," "Environmental Handprint," and "Ecological Handprint" to emphasize environmental and social meaning and importance, which goes beyond child's play and a literal dictionary definition. Likewise, I capitalize "Footprint" and "Ecological Footprint" when referring to the environmental resources consumed by a given action.

Our Environmental Handprints draws upon the genius and commitment of recent environmental authors. Amory Lovins, in *Reinventing Fire*, shows how we can reduce the cost to the environment of doing business while increasing profit margins. David Suzuki's *The Sacred Balance* explains how science and spirit align. In *The Quiet World* and *Rightful Heritage*, Douglas Brinkley chronicles how Theodore Roosevelt and Franklin Roosevelt protected whole landscapes from development. *The Carbon Farming Solution*, by Eric Toensmeier, renders carbon-sequestering agriculture practical—and, arguably, inevitable. In 2017, Paul Hawken gave us *Drawdown*, a broad, well-documented plan to reduce the carbon released into the atmosphere. I share the qualified optimism of their foundational work.

Our Environmental Handprints makes striving toward planetary health personal *and* collective.

This book is not a prescription or a plan. It *is* a call to take personal action. No particular technique will solve all our problems. Skills and personal interests of all sorts are needed. You can read a story to a child. You can volunteer for the Friends of [your favorite] River. You can donate to any of the environmental causes mentioned in these pages. You can lobby your county commissioner. You can drive a bus. All with a sense of mission rather than guilt.

I use "we" a lot to help us own the best efforts of humankind, past and present. Celebrating each other has power. John Muir, who championed national parks, is an ancestor to all who are inspired by his work (chapter 9). Bhutan's use of happiness to measure success establishes a precedent for all nations (chapter 18).

While climate change is the front-burner environmental issue of our times, *Our Environmental Handprints* addresses the broader swath of environmental issues using system interventions. Buying organic food helps build the soil. A plastic bag ban helps the ocean. An environmentally friendly house remodel reduces the Footprints of future occupants. Protecting the water in North Dakota brings indigenous ceremony, wisdom, and leadership to the environmental movement. Another way of defining Handprints is beneficial changes in how a system works.

I am a grandparent now. While I left behind the salary of full-time employment, I am not ready to let go of my responsibility to do what I can for the future nor my forty years of experience creating sustainability.

In 1975, my boss's boss catalyzed my journey to create (as I see it now) a lifetime Handprint. He invited me to coffee at his home to learn about the problems with nuclear power—cost overruns, the potential for creating bombs, and waste that stays radioactive for thousands of years. I still remember that first flip-chart page. It said, "Plutonium is bad for people." Over the next decade, I advocated for half a dozen statewide antinuclear ballot measures. I also quit my aerospace job and went back to college to shift from electrical to mechanical engineering. And I started work on my first commercial-scale solar project with a consulting engineering firm.

In 1980, a wholesale electric utility, Bonneville Power Administration, headquartered in Portland, Oregon, hired me as it began implementing its Pacific Northwest energy conservation mandate. For twenty-three years, I coordinated research and managed energy-efficiency programs. One program improved the efficiency of municipal water and wastewater treatment plants owned by Bandon, Oregon; Jackson Hole, Wyoming; Richland, Washington; and seventeen other Pacific Northwest cities.

Leaving my government job allowed me more time to create sustainability in the community. Portland's Village Building Convergence gave me hands-on experience with earth-friendly building (chapter 3). Advising Portland's Water Bureau and Green Investment Fund gave me civic experience. Organizational Development consulting has allowed me to help environmental organizations such as Cascadia Wild (citizen science), Ecology in Classrooms and Outdoors (education), and the Center for a Sustainable Today (video production).

In 2013, I presented "Our Environmental Handprint—The Good We Do" to an engineering conference, which gave me and my coauthors the opportunity to show how the Handprint complements the Footprint. We need both (see chapter 1).

My wife and I strive to live sustainably. We lived in a small (518-square-foot [48-square-meter]) house that allowed us to live on less income. Then we bought a fixer-upper and eco-remodeled it (chapter 12). And we've planted food forests (orchards and gardens) wherever we've lived.

HOW TO USE THIS BOOK

The purpose of this book is to empower you. You do not need to be an expert, wealthy, or part of an "in" crowd—unless you feel called to go there. You are welcome to choose your approach to healing our planet.

To appreciate the amazing breadth of our environmental progress and possibilities, read this book front to back. Every vignette holds the potential to enlighten and inspire.

To narrow your inquiry, consult the table of contents or the index. You will see many proven ways we can move toward sustainability. For instance, chapter 8, "Teach Ecology," explores environmental literacy from childhood through retirement, and a subsection highlights the experiential learning available from nature schools.

"Handprint opportunities," found at the end of each chapter, offer specific ways to make a difference. Some opportunities will be simple, like reading *The Man Who Planted Trees* to a child. Some require commitment, like creating an ecovillage.

Bring a notebook on this journey. Following each "Handprint Opportunities" section, a question prompts you to record your own reflections. Let the text and opportunities be catalysts. We often need to take small steps before acting on a big idea. My wife read about food forests and permaculture long before planting her first backyard food forest in the 1990s.

A special Handprint opportunity awaits you. In chapter 20, you will be invited to give yourself credit for environmental good that you have already done, your past Handprints. Then you will envision your personal intentions going forward, your future Handprints. In taking these simple steps, you move from awareness of possibilities to appreciating how *you* can heal our planet.

Refer to the appendixes to link your efforts to international sustainability movements. Appendix 1 links strategies to curtail our carbon emissions, as highlighted by Paul Hawken in *Drawdown*, to Handprint opportunities throughout *Our Environmental Handprints*. Appendix 2 links the seventeen UN Sustainable Development Goals to chapters in this book (see chapter 7).

Handprints are not about guilt. They are about love, creativity, and being called to serve. Join me in creating sustainability. Join me in responding to the challenge Greta Thunberg and her peers have set out for us.

I

CREATING SUSTAINABILITY

1

HANDPRINTS *AND* FOOTPRINTS

While visiting northern Arizona, I asked a Southern Paiute Indian friend to tell me the meaning of the handprint carved long ago into the black wall of a small canyon near his home. Instead of answering my question, he suggested that I lay my hand on the ancient petroglyph. Placing my fingers in the smooth indentations, I felt a peaceful connection with a living person, and so a people, and so their way of life. A sustainable way of life.

Over time you and I will also leave traces of what we care about. We are about to explore many ways we, as humans, create Handprints—that is, positive contributions to the world. Enjoy the journey. Feel that which will help us heal the planet.

THE HANDPRINT STORY

What is a Handprint? How did it evolve to become a symbol for sustainability? Why write a whole book about it?

The *Merriam-Webster.com Dictionary* simply defines a handprint as "an impression of a hand on a surface."[1] That is a literal definition, but we know there is more to it than that. Explicitly, a child's handprint in concrete is a record of a time when that child was young. Implicitly, it conveys the innocence of a child and the love of a parent who permitted or even arranged this imposition on the normal setting of concrete.

Whether we witness a child's handprint in concrete or touch the handprint carved in stone, we sense a positive, perhaps even healing, energy. In a business-oriented world, such a perspective seems like an indulgence in sentimentalism. But, as we shall see, the heart can move mountains.

The Centre for Environmental Education (CEE) in Ahmedabad, India, used the Handprint as a symbol for UNESCO's Fourth International Conference on Environmental Education in 2007. Not just any handprint, but that of a ten-year-old girl named Srija. The CEE website now tells us that a Handprint stands for "positive and tangible actions toward sustainability." They also suggest a deeper meaning as found in the phrases "joining hands" and "extending a helping hand in caring for the planet and all life on it."[2]

Greg Norris, an adjunct professor at Harvard University, tells us that a Handprint is "a contribution that causes positive change in the world—including reductions to your own or somebody else's footprint [the resources they consume]." He alludes to "creative handiwork" and "healing touch." Norris encourages students and companies to become NetPositive—that is, "putting more back into society, the environment, and the global economy than they take out."[3]

Rocky Rohwedder, a world-traveling professor emeritus at Sonoma State University, brings to the Handprint social and environmental justice perspectives. His 2016 e-book, *Ecological Handprints*, describes solar lighting, efficient cooking, and clean water projects in Asia, Africa, and Latin America.[4] Rohwedder asserts that successful Handprint projects are often affordable, local, women-empowered, digitally enhanced, and/or creatively financed. For example, in East Africa you can buy down the price of a solar electric system by paying for power as you use it—with money saved by not buying kerosene.

I came to the Handprint idea on my own before discovering other Handprint pioneers. My first Handprint publication, in 2009, simply stated that "an environmental handprint is the good one does for the world." My examples: Giving a child an environmental education gives her the opportunity to become an environmental heroine like Rachel Carson. And, buying bulk walnuts from the food cooperative saves packaging, uses plant-based protein, and supports a community-friendly institution. In a 2013 professional paper, my coauthors and I added the dimensions of collective Handprint (the result of many peoples' efforts), lifetime Handprint, emerging Handprint, and Handprint thinking (described later in this chapter) to the thickening stew of Handprint ideas.[5]

I also observed two valuable attributes of the Handprint. There is no theoretical limit to a Handprint's impact, and some Handprints can self-propagate. For example, removing a dam can allow a whole ecosystem to recover (chapter 16).

For these reasons, it is difficult to quantify the impact of a Handprint. (Advanced degrees could be earned doing so.) I prefer to count outcomes such as trees planted or toxic waste sites recovered. *At its finest, a Handprint changes the system itself.* An ordinance banning the use of plastic bags changes how we shop and reduces the flow of plastic to the ocean.

The Handprint, like a wrench, is a very useful tool with many variations. Serving a tasty plant-based meal tweaks the food industry with the exactness of needle-nosed pliers. Eco-remodeling your home applies a pipe wrench to the carbon and water Footprints of present and future occupants. Buying an electric car is like employing an adjustable open-end wrench to the auto industry. Enacting a constitutional amendment to end corporate personhood (not an easy task) can curb corporate excess with the effectiveness of vise grips.

The secret to the Handprint's effectiveness lies not in its definition or even in its calculation but rather in its potential to inspire and motivate life-affirming change. To that end, I will show in the coming pages how Handprints have yielded a vast array of underreported actions and outcomes that keep on giving. Thinking in terms of Handprints offers *each of us* whole realms of planet-healing opportunities.

PLANTING TREES

Planting a tree is a simple and tangible way we can create a Handprint. Pick the spot thoughtfully, water the seedling, and the tree grows pretty much on its own. As it gets older, it will provide perches for birds, erosion control for hillsides, shade for a hot summer's day, and beauty for our emotional well-being. Some trees will provide fruit and seeds for animals of all sorts, including us. All trees sequester carbon from the atmosphere.

How much carbon does a tree sequester? That number depends on the type of tree, where it is planted, and how old it is. One answer calculated by the University of North Carolina extension is forty-eight pounds (twenty-two kilograms) of carbon dioxide sequestered per year, or roughly a ton (English or metric) by the time the tree is forty years old.[6] One tree will not reverse global warming, and we need results in fewer than forty years. But planting a lot of trees certainly helps.

In Kenya, streams were drying up and people had to walk increasingly long distances for firewood and fence posts. In 1977, Wangari Maathai started what became known as the Green Belt Movement. The intent was to plant trees while empowering women. Since then, over fifty-one million trees have been planted throughout the Congo Basin. For this work, Wangari Maathai was awarded the 2004 Nobel Peace Prize.[7]

Inspired by Maathai, the United Nations started the Billion Tree Campaign in 2006. It sought to reforest areas deforested by refugees while employing those same refugees.[8] After exceeding their goal many times over, the United Nations handed over the Billion Tree Campaign to the Plant for the Planet Foundation, with its emphasis on involving children. Over seventy thousand children around the world have earned the designation Ambassador for Climate Justice![9] The plant count continues—nearly fifteen billion trees as of 2019.[10]

Thomas Crowther and an international team calculated that there are just over 3 trillion trees in the world—1.39 trillion in the tropics and subtropics, 0.61 trillion in the temperate forests of the middle latitudes, and 0.74 trillion in the boreal forests in the far north. According to Crowther, we cut down about fifteen billion trees a year and plant about five billion a year.[11] Reversing those rates would be a worthy goal.

That is exactly what Felix Finkbeiner intends to do. As a nine-year-old, he proposed to his fourth-grade class that children plant one million trees in every country. His motto: "stop talking, start planting." A decade later, this vision has evolved into the Trillion Tree Campaign, sponsored in part by Prince Albert II of Monaco and Klaus Töpfer, director of the United Nations Environment Programme from 1998 to 2006.[12]

These initiatives have a robust precedent. During the 1930s, tree planting helped the United States respond to the environmental degradation of the Dust Bowl and the destitution of the Great Depression. Under-employed men planted almost three billion trees, working with the Civilian Conservation Corps and other federal programs.[13]

When you plant a tree, you are not alone.

THE ECOLOGICAL FOOTPRINT

The Handprint's elder brother is the Footprint. The Ecological Footprint is a measure of the resources we require for our daily living. It is not, as popular sentiment implies, the same as environmental damage.

Our Ecological Footprint, published in 1996 by Mathis Wackernagel and William Rees, brilliantly asserted that the earth has an amazing but finite capacity to provide energy and process waste. This truth framed our ecological situation in terms of the carrying capacity of the land. The authors showed how we, along with our cities and products, are all part of a system we call Earth.[14] The authors developed a calculus for determining how much of the mix of productive land that is actually available (measured in global acres or hectares, gha) it takes to meet a given level of consumption. From that we can determine how many Earths it would take if everyone consumed similarly.

The Ecological Footprint has definitely changed how people think. Minimizing one's Footprint, or how much energy we consume and how much waste we produce, is as much a mantra of the eco-conscious citizen as "reduce, reuse, recycle."

The Global Footprint Network tracks how the world is doing. It specifically considers arable land, grazing land for animal products, fishing in "productive" (typically coastal) waters, forest for wood products, forest for carbon sequestration, and "built-up land" where we live in cities. Biologically nonproductive lands such as deserts and Antarctica are not included in this global accounting.

About 4.2 acres (1.7 gha) of productive land (and water) is available per person on this planet. I only get to occupy a small portion of that. We in the United States use resources at the rate of 20 acres (8.1 gha) per person (2016 data). If everyone consumed as much as the average American, we would need about 4.8 Earths.[15]

Earth Overshoot Day "marks the date when humanity has used more from nature than our planet can renew in the entire year." That day was September 23, 2000. By 2019, Earth Overshoot Day had moved back to July 29. Put another way, we have been diminishing our children's ability to survive and prosper since 1969—when we first exceeded our global carrying capacity.[16]

FOOTPRINT *AND* HANDPRINT THINKING

The Handprint, for me at least, was born out of my perceived inability to reduce my Footprint. Family, employment, and entertainment compete for my attention. Tali Sharot, in *The Influential Mind*, reports that "we are more likely to execute an action when we are anticipating something good than when we are anticipating something bad."[17] We avoid getting overwhelmed when we think in terms of Handprints.

However, a coauthor of my Handprint professional paper—my wife, Willow—impressed upon me an important point. Doing a lot of good does not absolve us from being responsible for the resources we use. A difficult conversation, two days before a midnight submission deadline, led us to see how Footprint thinking and Handprint thinking are *complementary* rather than exclusive opposites. The Handprint is built on the foundation of the Footprint. A person needs hands *and* feet.

Footprint thinking helps us track data. The average car on the road in 2011 traveled 11,400 miles at 26 miles to the gallon to release 5.2 tons (4.7 metric tons) of carbon dioxide from its tailpipe.[18] Handprint thinking highlights changes in the system, such as banning plastic bags in eight states.[19] Recovering the Rhine, Thames, and Cuyahoga Rivers are Handprints (chapter 16).

Footprint thinking stops mountaintop-removal coal mining. Handprint thinking fosters solar and wind power.

Footprint thinking emphasizes limits, inviting us to "reduce, reuse, and recycle." Handprint thinking looks for possibilities: "recover, reallocate, and re-envision."

Footprint thinking acknowledges the carbon released into the atmosphere from cooking with wood. Handprint thinking delivers more efficient stoves, improving indoor air quality and the lives of families.

Footprint thinking inclines me toward an earth-centric perspective. Handprint thinking embraces environmental justice and the needs of the developing world (chapters 6 and 7).

Footprint thinking shows us how our consumption moves Earth Overshoot Day earlier and earlier each year. Handprint thinking takes fate in our own hands and helps us again live within our means.[20]

We need Footprint *and* Handprint thinking!

VOLUNTARY CARBON OFFSETS

Footprint and Handprint thinking come together naturally with voluntary carbon offsets. The carbon released by my home, my transportation, my wedding, my business, or my conference can be "offset" by supporting projects that reduce carbon emissions elsewhere—Handprints.

Buying carbon offsets typically involves three steps: Assess your own carbon footprint. Choose what kind of project—Handprint—you want to support. Then choose how much you want to give to support that project based on the size of your carbon footprint. Carbon Fund, Conservation International, and Native Energy make this process easy through their websites. Corporations and colleges can also contract BEF, ClearSky, or 3 Degrees to arrange large-scale offset initiatives.

Through carbon offsets, we help bring some challenging projects to fruition—capturing methane emissions from an abandoned coal mine in Pennsylvania; protecting rainforests in Belize; building a remote wind farm in Alaska; and developing regenerative rice farming in Arkansas, Mississippi, and California.[21] Biologists, engineers, and entrepreneurs have been developing the technology—all independently verified. You and I help with the financing.

When I flew from Portland to San Francisco to speak at a conference, the organizers paid an extra seven dollars to offset the carbon released by my plane. That small chunk of change went to a Vermont-based carbon-offsetting company called Native Energy, which partnered with Ghanapreneurs LLC to provide water filters to families in Accra, Ghana.[22]

How does this reduce carbon? The Ghana Clean Water Project provides blue plastic water filters, about the height of a four-year-old, to families who do not have access to clean drinking water. Otherwise, those with the financial means would be buying wood or charcoal to boil water before drinking, causing air pollution, endangering health, and releasing carbon into the atmosphere. Over ten years, the project will reduce carbon emissions by seventy-seven thousand tons (seventy thousand metric tons). This is a complex arrangement, but Viability Africa, an independent organization, uses a United Nations protocol to verify that the project works as claimed.

In 2015, Unitarian Universalists from around the country offset travel to their annual meeting in Portland, Oregon. Part of the registration fee went to the Carbon Fund, which funded electrical hookups at truck stops to reduce layover idling. A sustainable event support company called Meet Green made the necessary arrangements.

I used Conservation International's calculator to calculate my personal carbon footprint.[23] It took into account a low-meat diet, recycling, use of public transportation in lieu of a car, and . . . a flight across the country to Pittsburgh to visit my dad. The website dutifully added four tons of carbon (requiring fifty-seven trees) to my calculated total, accounting for "goods and services, e.g., clothing, furniture, appliances, entertainment, personal care, health care and education." That is a way of allocating societal carbon releases that come with living in the United States. Thus, my total carbon footprint is twelve tons of CO_2e/year. (For consistency, carbon footprints are reported in terms of tons [or metric tons] of the equivalent carbon dioxide impact on global warming.)

My offset price: $144. According to Conservation International, it will take 170 trees to offset my 12 tons of CO_2 (at the rate of 0.07 tons per tree).

Put another way, I am helping protect Madagascar's Ambositra-Vondrozo Forest Corridor, where "over 800 species of plants and 300 species of animals have been identified in these forests, including 17 species of lemur."[24] Add to that the 670,000 people who call that area home.

When I suggested paying for just the airfare portion of my offset, Willow said, "I think we ought to pay our full rent on the planet and not try to cheat the landlord."

HANDPRINT OPPORTUNITIES

1.1 Introduce children to trees. Read *The Man Who Planted Trees* by Jean Giono to children. Name a tree. The Friendly Tree, a sugar maple, lived behind my Pennsylvania home.

1.2 Buy 100 percent recycled printer paper. Rite Aid sells it.

1.3 Move to online bill paying. Save trees even when you are distracted.

1.4 Switch to double-sided (duplex) printing.

1.5 Switch to reading e-books from Apple, Kindle, Kobo, or Nook.

1.6 Plant trees. Leave a legacy. Honor the departed. Recover the land.

1.7 Donate to plant trees. Check out the Arbor Day Foundation Rainforest Legacy Program or Plant-for-the-Planet.org.

1.8 Volunteer for your local arboretum or botanical garden. Lead tree tours. Help others appreciate the miracle of trees and other flora.

1.9 Endow the Tree Fund to sponsor research and education relating to urban forestry. Example 2015 grant project: "Quantifying the Cooling Effectiveness of Urban Trees in Relation to Their Growth," by Mohammad Asrafur Rahman, Technical University of Munich, Germany.[25]

1.10 Research how trees sequester carbon. We could use a PhD and a book on that subject. How does the rate of sequestration vary by age of the tree?

1.11 Use the online Ecological Footprint Calculator to quickly assess your demands on nature. Try entering different eating habits, housing and transportation. Let the possibilities inspire you. Invite a friend to do the same.[26]

1.12 Offset your carbon footprint. Variations: offset your business travel, your family's air travel, or your whole year's carbon emissions. Check out the Carbon Fund, Conservation International, Native Energy, and Nature Conservancy. Terrapass offers a bundle option for your wedding and a monthly subscription to support renewable energy.[27]

1.13 Develop a carbon offset policy at work. Will your company offset its air travel? Employee commutes? The whole company's carbon footprint?

IN YOUR JOURNAL

I suggest keeping a journal to record *your* next steps to help heal our planet.

2

LITTLE THINGS ADD UP

A s Willow and I walked through our wonderfully reclaimed wildlife area, she listened thoughtfully as I talked about being a hero in one's own story and how people in ecovillages create Handprints. The area used to be home to decaying houses, a used tire store, and discarded refuse. Not many people had lived in the area because the nearby Johnson Creek flooded regularly. Now, after the removal of debris and re-meandering the stream, the area is a home for beaver, deer, mice, and hawks. All within the city limits. Eventually, Willow quietly interrupted me with two words: "Little things."

WILLOW'S APPROACH TO SUSTAINABILITY

Three questions:

1. What things do I do most often?
2. What are the products I use most often?
3. Where do things go when I'm done with them?

Three challenges:

1. Replace a disposable with a reusable.
2. Replace toxic with nontoxic.
3. Look for durable and/or biodegradable alternatives.

By switching to environmentally savvy products and practices to address our daily needs, we can both reduce our Footprints and increase our Handprints.

CLEANING PRODUCTS

If you use less stuff, especially toxic stuff, you reduce your Footprint. If you buy more good stuff on a regular basis, you create an eco-industry-supporting Handprint. Rather than holding on to unsustainable habits, let's just recognize that some things are really worth doing.

Many of the cleaning and personal care products available in the supermarket contain petroleum products and secret ingredients. The Environmental Working Group recommends making your own cleaners with white vinegar, lemon juice, baking soda, or washing soda. I found that water with vinegar and a little bit of dishwashing liquid works great to clean our windows. Baking soda and a little effort cleans the tub. The Environmental Working Group recommends vinegar and salt as a grease-cutting surface cleaner.[1]

Top-quality microfiber cloths, such as those available from FlyLady and Norwex, enhance almost any cleaning experience. With this twenty-first-century technology, you can often get the job done with plain water, but some microfiber cloths perform better than others.

A wide variety of ingredients can be used to make soap. Many artisan soap makers avoid both petroleum- and animal-based ingredients.[2]

Biokleen has built a business out of providing environmentally friendly products tailored for particular challenges. We use their laundry liquid for our clothes and their Bac-Out when a pet has an accident. Skeptical though I was, their Drain Care, with live enzyme-producing cultures, worked fine with a couple treatments for a slow-moving drain. All three products note on their labels, "No materials listed by the ACGIH (American Conference of Governmental Industrial Hygienists) as hazardous."

You can get treeless toilet paper made out of bamboo, a fast-growing renewable resource, and sugarcane stalks, a waste product. Brands include Emerald, NaturEZway, NooTrees, Rebel Green, and Caboo.[3] Again, performance may vary. Who Gives a Crap, capitalized with $50,000 of Indiegogo crowdfunding, offers both bamboo and recycled-paper toilet paper. Green Forest and Seventh Generation brands also deliver recycled paper products with 50 percent or more *post-consumer* content.[4]

INTELLIGENT CLOTHING

Much of the clothing available today contains petroleum-based fibers. Nylon, acrylic (Orlon), and polyester are all petroleum-based materials that use up the earth's resources and create additional waste in their production. We can do better. There are many other materials and fibers that are more environmentally friendly and will help increase your Handprint!

Bamboo fiber is soft, inexpensive, and antiallergenic. The bamboo plant removes five times as much greenhouse gas as most trees and yields ten times the fiber per acre as cotton—without the need for replanting or pesticides. It can withstand both drought and flood. Besides clothing, bamboo is made into sheets and towels.[5]

Industrial hemp is also extraordinarily versatile and durable. It is made into designer jeans, dresses, and sandals as well as rope, ship sails, paper, and wallboard. Hemp produces four times as much fiber per acre as forests. Pre–World War II, many farmers raised hemp, but that was legally curtailed largely through the efforts of the timber industry in the United States, which saw hemp as a competitor for making paper. Industrial hemp is low in THC (tetrahydrocanabinol), so its association with high-THC cannabis (marijuana) was a political rather than practical issue. Buying *organic* hemp clothing helps revive the industrial hemp industry with eco-values.[6]

Ramie, also called China grass, is a natural fiber from the nettle family that may be used instead of, or blended with, cotton. It was used to wrap Egyptian mummies. Its strong, washable fibers are made into suits, skirts, jackets, dresses, shirts, pants, handkerchiefs, draperies, canvas, and fire hoses. It is a premium fabric because of its relatively high production cost.[7]

Organic cotton is much better for the environment than "conventional" cotton.

BENEFITS OF ORGANIC COTTON COMPARED TO CONVENTIONAL COTTON

- Twenty-six percent less (over)fertilization due to soil protection procedures to prevent erosion.
- Forty-six percent less carbon dioxide equivalent (CO_2e) released into the atmosphere due to reduced tractor operations, fertilizers, pesticides, and irrigation.

- Ninety-one percent less water use due to improved irrigation methods.
- One hundred percent reduction of toxic chemicals used for pesticides. That is, *none*.
- One hundred percent reduction of genetically modified seeds. That is, *none*.

Source: Textile Exchange life cycle assessment conducted in 2014, Textile Exchange, "Quick Guide to Organic Cotton," Textile Exchange, June 7, 2017, https://store.textileexchange.org/product/quick-guide-to-organic-cotton.

Considering the entire clothing market in 2015, the Textile Exchange found that organic cotton sales saved 57 billion gallons (218 billion liters) of water and reduced carbon emissions by 101 million tons of CO_2e (92.5 million metric tons).[8] Manufacturers that offer organic cotton in their product lines include Gucci, Indigenous, Nike, REI, Patagonia, and Under the Canopy.[9]

Recycled polyester (rPET) clothing provides a market for recycled plastics and has about two-thirds the carbon footprint of virgin polyester. This is significant given that nearly half the world's clothing is made of polyester. However, at the end of the product cycle, organic fibers that naturally break down leave less of a burden on the environment.[10]

Patagonia, based in California, helped pioneer the concept of a benefits-driven company. Patagonia sources 75 percent of its outdoor apparel from "environmentally preferred sources" such as recycling and organic suppliers. All of Patagonia's suppliers are made transparent on its website. "The Board includes independent members to represent interests of Community and Environment."[11]

Patagonia also hosts the Worn Wear brand of clothing made in Los Angeles from recrafted trade-in products.[12] This is one example of a burgeoning industry of upcycled clothing sold online. Thankful Rose makes upcycled women's clothing in its Salam, Oregon, shop.[13] Their one-of-a-kind tunics, skirts, sweaters, scarves, and wrist warmers are sold online through the Etsy website. There you will also find eco-friendly treasures from dozens of other crafters of classy clothing.[14]

ThredUp claims to be the "largest online consignment and thrift shop." Their website conveys their game-changing Handprint intent: "Make a real fashion statement—choose used." "Think secondhand first." "The future is secondhand."

ThredUp makes it easy to mail in used clothing for store credit or to trigger a five-dollar donation to charity. The times are changing. Fifty-eight million women bought secondhand in 2018, up from forty-four million in 2017.[15]

Read more about the resale industry in chapter 18.

ELIMINATING DISPOSABLES

Americans typically leave behind a lot of waste. We are constantly tempted to suspend our environmental values when visiting friends, rushing to take care of business, and working away from home. Whenever and wherever you repeatedly use a disposable item, you have the opportunity to make an eco-upgrade Handprint. Willow asks of herself, "What do I use every day that is disposable? How can I replace it with reusables?" She now has answers to those questions.

Lightweight, packable shopping bags are available from ChicoBag and Baggu. Weather-ready garments can be kept in their own stuff pockets. Check out packable rain jackets and rain pants at Columbia Sportswear, Outdoor Research, and REI. With these in your purse, briefcase, messenger bag, or day pack, you can be eco-ready wherever you are.

You can carry what you need to eat a simple meal. Packing your own spoon, fork, or spork saves the landfill and atmosphere one plastic utensil every time you use it. Likewise, having your own bowl eliminates the need for paper and plastic plates. Wrap these things in a bandana to provide a reusable napkin.

A refillable water bottle avoids the need to redeem and recycle single-use plastic water bottles. A smoothie shaker can carry a homemade smoothie and later serve as a water bottle. A collapsible Hydaway Bottle can be compactly kept in your purse, pocket, or day pack when not in use.

For day trips, Willow carries a stainless-steel PlanetBox filled with eco-healthy food (see chapter 11). We do not *have* to eat fast food from drive-through disposables.

These little things add up to real Handprints!

HANDPRINT OPPORTUNITIES

2.1 Buy plant-based soap. Some members of the Handcrafted Soap and Cosmetic Guild make plant-based soap. Check for a soap maker in your state.[16] Or, check your local food co-op for artisan soaps.

2.2 Buy earth-friendly toilet paper. Look for bamboo, sugarcane stalk, or recycled paper with high post-consumer content.

2.3 Give cloth napkins and reusable shopping bags as gifts. Influencing others is a Handprint.

2.4 Carry and use a reusable spoon, fork, or spork. And a reusable bowl. Wrap them in a bandanna, which can serve as a napkin and a towel.

2.5 Buy bamboo, ramie, organic hemp, or organic cotton clothing.

2.6 Buy classy used clothing. Check out the Etsy and ThredUp websites.

IN YOUR JOURNAL

How will *you* change your routine way of doing things to create a personal Handprint?

II

ALLYING WITH HUMANITY

3

NURTURE COMMUNITY

At our neighborhood street party a few years ago, one neighbor volunteered a front yard, two neighbors offered barbecues, and others brought food. Someone collected names and phone numbers. We now share resources and come to each other's aid. The musician family down the street loaned me their car for a couple days to help my wife through a health crisis. The neighbor who works on an organic farm harvests our bamboo. Labels like Republican, Democrat, and radical fell away.

If neighbors know each other, they need fewer resources and they will be better stewards. They will be less lonely and more prepared for disasters. And they will be more empowered to create Handprints.

COMMUNITY SHARING

On Monday mornings at Portland's Hollywood Senior Center, volunteers distribute vegetables left over from the local farmers market. On Tuesday mornings, shelf items that are near their expiration dates from the local New Seasons grocery store are shared with seniors. People talk pleasantly with each other while waiting in line as much as an hour before the door opens.

While not billing itself as an environmental hotbed, the senior center sure has a lot to offer. A large rack loaded with quality magazines offers visitors free reading. A bookshelf facilitates the exchange of novels. Over in one corner is the free table, where people leave puzzles and sweaters that someone else might use. A bulletin board lists activities for the day, ranging from yoga to cooking to ukulele.

Now let us count the benefits: reduced food waste, saving trees, and sharing skills. That is in addition to helping seniors, promoting health, and building community. And one more thing: the senior center rents space to a congregation on Sundays. Benefit: greater utilization of resources and financial support for all the other senior center activities. Besides, the congregation feeds the seniors dinner every Thanksgiving.

Other senior centers, community centers, businesses, faith communities, and nonprofits have similar community Handprints.

In the same spirit of free exchange, Rebuilding Together Central Ohio, based in Columbus, runs a tool lending library, including frequently used lawn mowers. There is no charge to users. Some of the tools have been donated by Home Depot.[1] Tool lending libraries can be found in Erie, Pennsylvania; Minneapolis, Minnesota; Oakland, California; and West Seattle, Washington, among other places.

Another model of community sharing is the time bank, sometimes known as a skill exchange. Members of the Toe River Skill Exchange in Burnsville, North Carolina, "pay it forward." I help another member with my skill and time, then log in my hours. Later I can request help, perhaps of a different sort, from another member, hour for hour. Everyone's time is valued equally.[2] The concept is described in Edgar Kahn's book *No More Throw-Away People*, published in 2000.[3] The Directory of Time Banks lists Sebastopol, California; Saint Petersburg, Florida; Meadville, Pennsylvania; Gumi, South Korea, and over a hundred other time banks.[4]

REPAIR CAFÉS

Another movement that is quietly growing, beneath the attention of the media, includes folks who repair your clothes, small appliances, and bicycles for free. Martine Postma started the first Repair Café in Amsterdam in 2007. She wanted to find a way to counter the throwaway mentality that affects both product makers and product owners. The idea of gathering a few skilled people in one place on one day and offering their services to whoever comes—for *free*—took off. There are now more than 1,500 Repair Cafés convening regularly around the world.

I took a kid's scooter on my first visit to a Repair Café. Its handle had locked down in such a way that I could not get it loose. I took it to the pair of bicycle repair people. Both of them studied why it had gotten locked down. Turns out the lever that allows you to extend the handle was caught under the tail bar. A little loosening, twisting, and retightening, and that problem was solved—in under five minutes. At my next Repair Café, a master of the sewing machine repaired a hole in my robe, but the knife sharpeners chose not to take on the complex blade of my pruning saw.

Repair Cafés claim a 70 percent success rate in fixing the things people bring to them. For those items that cannot be fixed on the spot, a referral is often made to a working professional. Kudos to the institution that Martine Postma founded, a very nice Handprint. Kudos to the people who donate their time to create four to eight modest Handprints, per craftsperson, per evening. Kudos to me and thousands of participants for not throwing away repairable stuff.

ECOVILLAGES

Residents of intentional communities share resources and align themselves with shared values. This kind of cooperation naturally reduces each member's Ecological Footprint. Today, there are some 3,000 intentional communities in the United States, 1,200 of them in the directory of the Fellowship for Intentional Community.[5] I met Willow at Shannon Farm, in Virginia, one of those communities.

Members of ecovillages, a subset of intentional communities, include in their commitment a focus on environmental values.

Dancing Rabbit Ecovillage, in rural northeast Missouri, is one example. It was founded as a 280-acre land trust in 1997. Dancing Rabbit offers a visitor program, workshops and webinars to explain natural and green construction, alternative currency, renewable energy, inner sustainability and communication skills, and village decision making and governance.[6] Every member agrees to follow a set of ecological covenants.

ECOLOGICAL COVENANTS
OF DANCING RABBIT ECOVILLAGE

- No personal motor vehicles on the property.
- No use of fossil fuels for vehicles, space conditioning, or water heating.
- All gardening and landscaping shall follow organic procedures.
- Imported electricity must be balanced with sustainably generated on-site energy.

- Lumber for construction shall be reused/reclaimed, locally harvested, or certified as sustainably harvested.
- Organic and recyclable materials shall be reclaimed.

Source: "Here Are the Ecological Covenants People Who Live at Dancing Rabbit Agree to Adhere To," Dancing Rabbit Ecovillage, accessed April 18, 2020, https://www.dancingrabbit.org/about-dancing-rabbit-ecovillage/vision/ecological-covenants.

EcoVillage at Ithaca, in New York, has 175 acres, 55 of which are in a permanent conservation easement (no buildings allowed).[7] When fully built out, EcoVillage at Ithaca will have one hundred energy-efficient homes in three neighborhoods and office space for entrepreneurs. People in the surrounding area subscribe to buy produce from two on-site organic farms, operated by residents on land leased from the community for the cost of taxes. This demonstration of community-supported agriculture (CSA) is shared with students from Ithaca College. Theirs is a cohousing model where individuals can sell or rent their homes through the village website.[8]

Seventeen residents of Cincinnati, Ohio, met in 2004 to form the Enright Ridge Urban Ecovillage. Over time they purchased and eco-refurbished several foreclosed homes with the help of a foundation. They also started a CSA using backyard gardens and employing farmers. About 40 percent of the ninety homes on the three-quarter-mile dead-end street are directly involved. There is no opposition to the ecovillage because it is a great place to live.[9]

Ole and Maitri Ersson, in Portland, Oregon, resurrected a complex once used as a gathering place for drug users to create the Kailash Ecovillage. This two-acre ecovillage uses a rental model rather than a cohousing model to allow people with lesser means to join. The complex has thirty rental units and two hostel-style shared living spaces. On-site rain gardens collect water from the main building's roof and the remaining parking area. Residents de-paved twelve spaces of the original parking lot. Food for frequent vegan community gatherings is provided by forty-six individual garden plots plus community-gardened space.[10] (Potlucks of any sort nurture community.)

VILLAGE BUILDING CONVERGENCE

Since 2001, City Repair has hosted the Village Building Convergence (VBC) in and around Portland, Oregon. The fundamental work of the VBC is "placemaking"—"the act of creating a shared vision based on a community's needs and assets, culture and history, local climate and topography." The act of gathering together with other people in a given area creates psychological ownership. The creation of a commons helps break down the socially sterile grid structure that pervades modern cities.[11]

In 2007, around Memorial Day, Willow and I invited VBC participants to help us replace our lawn with a food forest. Over several days, people came to spread cardboard and wood chips over the lawn and plant berry bushes and fruit trees. In the evenings, builders, neighbors, and friends convened at a central venue to share food, hear eco-talks, and dance. Three neighbors have since planted gardens in their front yards.

Placemaking occurs wherever neighbors create a friendly community commons. Building a cob building is a common VBC activity. Cob is straw and clay mixed together to form walls and benches; it's an eco-friendly and labor-intensive process. Painting a bright pattern where neighborhood streets meet, known as "intersection repair," encourages folks to hang out on corners and hold street fairs. Conversion of an old building into a community center transforms dereliction into connection. Painting a mural with many hands can create a sense of shared pride and responsibility for a neglected space. A street-side book-exchange kiosk encourages unexpected positive interactions. Portland, Oregon, officials now use placemaking language to describe new bus and light rail stops.

Using placemaking principles, the Project for Public Spaces has helped create Detroit's Campus Martius Park; Chicago's Washington Park Rain Garden; Cornog Plaza in Fort Meyers, Florida; Foro Lindbergh (Parque México) in Mexico City; and hundreds of other community-nurturing locations in fifty countries.[12]

COMMUNITY-FRIENDLY BUSINESSES

Healthy communities need support from businesses.

On New Year's Eve in both Charlottesville, Virginia, and Victoria, British Columbia, shops around the city open their doors to everyone. First Night, as it is called, offers free music, fun for children, a reason to come to town for visitors, and fireworks at midnight.

Sandy, Oregon, businesses agree to host trick-or-treaters on the Saturday before Halloween. The kids look forward to the day. Parents and children—future customers—come from miles around to join the festivities.

Benefit corporations often have a strong community-supporting mission (see chapter 18).

COMMUNITY RIGHTS

Regrettably, there are still businesses that demonstrate disregard for local communities and their values. Parts of rural America have been referred to as "environmental sacrifice zones." The Community Environmental Legal Defense Fund (CELDF) started as a law firm that used existing laws to challenge incinerators and waste dumps, but their clients were often defeated by a legal system designed to accommodate unsustainable growth, extraction, and development. Now CELDF proactively helps communities claim rights through ordinances.[13] Sometimes a community needs to change the rules.

Within community-rights ordinances are statements such as, "All residents, natural communities and ecosystems in the Town of Wales [New York] possess a fundamental and inalienable right to maintain the sustainable access, use, consumption, and preservation of water drawn from natural water systems, springs and wells that provide water necessary to sustain life within the Town."[14]

In 2009, Shapleigh and Newfield, Maine, passed ordinances to stop Nestlé from extracting water for bottling from the Vernon Walker Wildlife Preserve. Nestlé withdrew all twenty-three test wells.[15]

Jackson and Josephine Counties in Oregon banned the growing of GMO crops in May 2014.[16] The Oregon legislature negated Josephine County's ban but not the bans of California's Marin and Mendocino Counties and Hawaii's Kauai and Hawaii Counties.[17]

Halifax, Virginia, claimed the right to sue corporations for bodily violation by toxic materials from proposed uranium mining in nearby Southern Virginia counties.

Pittsburgh, Pennsylvania, was the first city to ban fracking within its city limits in 2010, and other communities—in Pennsylvania, New York, Maryland, Ohio, and New Mexico—followed. The April 29, 2013, Mora County, New Mexico, ordinance reads, "It shall be unlawful for any corporation to engage in the extraction of oil, natural gas, or other hydrocarbons within Mora County [and] to extract water from any surface or subsurface source, within Mora County, for the use in extraction of subsurface oil, natural gas or other hydrocarbons."[18]

In September 2013, Sangerville, Maine, enacted the nation's first "Community Bill of Rights Ordinance to protect a municipality from infrastructure projects without the consent of the voters." The "East-West Corridor" is a proposed 220-mile "corporate highway" intended to transport Canadian extractive materials (e.g., tar sand bitumen) across Maine to the Atlantic Ocean. Now Sangerville has a say.[19]

These actions, in some two hundred communities, serve as the foundation of a grassroots community rights movement. Community rights networks foster citizen-led democracy within Colorado, New Hampshire, New Mexico, Pennsylvania, Ohio, Oregon, and Washington. The National Community Rights Network is sponsoring an amendment to the US Constitution to protect community rights.[20] The *Community Rights Do-It-Yourself Guide to Lawmakers*, published by the CELDF, describes the basic legal challenges to community rights and how to overcome them.[21]

HANDPRINT OPPORTUNITIES

3.1 Start a free table. Settings can include cooperatives, schools, community centers, and social gatherings. One item lots of people can use is glass containers. Some monitoring and periodic sweeping is needed. Indoors may be best.

3.2 Support your local community center, or anyplace people can find healthy community with environmentally friendly services. Volunteer. Donate. Spread the word.

3.3 Participate in or start a tool lending library.

3.4 Participate in or start a time bank.

3.5 Create a kiosk in front of your house. Use it to exchange books, magazines, poems, and information with neighbors.

3.6 Arrange a street party along your main street or in your neighborhood.

3.7 Paint an intersection. Invite a local artist to create the design. Jostle local bureaucracies to get permission.

3.8 Paint a community mural. Offer the graffiti-prone wall of your business. Coordinate with a local school. Perhaps your town has a theme that can play out in several places. Check out how murals revitalized Chemainus, British Columbia.[22]

3.9 Plan a community-wide event. Think Meadville, Pennsylvania's, Veterans Day parade.

3.10 Participate in a Repair Café. If you are distant from one of their 1,900+ host groups, start one.[23]

3.11 Create an ecovillage out of your neighborhood.

3.12 Buy locally made products. Keep the environmental cost of transportation down while boosting the local economy.

3.13 Run for office if your commissioners, counselors, and mayors lack sufficient sensitivity. You can educate people even if you do not win.

3.14 Confront inappropriate corporate incursions in your community. Case in point: on May 17, 2016, local activists inspired voters of Hood River County, Oregon, to pass Measure 14-55, stopping Nestlé from bottling water from Cascade Locks' Oxbow Springs.[24]

IN YOUR JOURNAL

How will *you* create a community Handprint?

4

GIVE GENEROUSLY

One of the functions of community is to foster generosity. Yet, we also have the capacity to be generous with people we have not met.

THE GIVEAWAY

The giveaway helps distribute wealth and good fortune in a sustainable community. In Native American traditions, if a couple gets married, the family does a giveaway. If you recover from a major illness, you do a giveaway. It may be considered part of the healing ceremony that precedes it. We might prepare a year ahead of the giveaway by making things for it. We can look around and let go of some nice things or things we no longer need with a sense of thankfulness. Or we can make or buy things that people need, like canned food, knitted clothing, and tools. Think of the giveaway as art.

One aspect of the giveaway is sometimes forgotten. It is important for the community to accept all that is given (except toxic waste!). We take the seemingly useless trinkets too, even if we will soon donate them. This allows the giver to move forward cleanly.

What a model for a sustainable world!

GLEANING

Our plum tree yields far more plums than we can eat. We invite friends to collect buckets of plums for jellies and other recipes. Sometimes I take a child's wagon filled with surplus garden produce around the neighborhood—great for building community and reducing carbon emissions.

In 1979, the Society of St. Andrew was founded by two ministers and their families on a farm in Big Island, Virginia. They formed an intentional community dedicated to "helping resolve the problem of world hunger." One of their first projects was to deliver a tractor-trailer load of surplus sweet potatoes to the Central Virginia Food Bank in Richmond, Virginia.[1] The Society of St. Andrew now distributes food from over 550 farms to some 1,400 feeding agencies in a dozen eastern states.[2] Volunteers glean leftover crops for the Potato and Produce Project nationwide.

Urban Gleaners, a Portland, Oregon, organization, picks up surplus food from caterers and markets for distribution to relief agencies, schools, and "summer free farmers markets" in low-income school districts.[3]

According to Feeding America, food bank pantries and meal programs secure and distribute 4.3 billion meals a year in the United States.[4] In 2019, Trader Joe's stores donated seventy-eight million pounds of food to combat hunger.[5]

CROWDFUNDING

Brock Ketcher, Naomi Ketcher, and Luke Miner started YouCaring to help people pay off student loans. They assert that the world's most valuable untapped resource is compassion. Since 2011, YouCaring has raised "more than $1 billion in support of individuals, families, communities, and nonprofits looking to save lives, remember loved ones, recover from disasters, and fund humanitarian efforts." YouCaring has since become part of GoFundMe, the largest free crowdfunding platform.[6]

Kickstarter is a benefit corporation committed to making a positive social impact. It focuses on creative nonprofit projects. Since 2009, the organization has kickstarted over 170,000 projects with $4.6 billion from seventeen million people.[7]

Indiegogo brings crowdfunding power to international (for-profit) entrepreneurs. Since 2008, Indiegogo has raised over $1.6 billion "for hundreds of thousands of companies—from nearly 10 billion backers," according to founder Slava Rubin.[8]

Imagine turning our endless highways into an ecological asset by paving them with solar cells! Indiegogo helped bring this dream of Scott and Julie Brusaw into the realm of reality. Having paved an asphalt parking lot with solar collectors using their own money, they needed capital to build a company and an engineered technology that could be run over by thousands of eighteen-wheelers. Over fifty thousand individuals gave a total of

$2,287,809 for Solar Roadways' Indiegogo crowdfunding campaign.[9] That includes ten dollars from me!

ONLINE GIFTING PLATFORMS

An elegant way to keep our waste out of the landfill is to give it to people who can use it.

Freecycle helps us do just that.[10] You see what others want to give away, and you announce what you have to give away. My hard-to-find cassette player came from Freecycle, as did an almost-new water heater. We gave away our old television with a built-in videocassette recorder. I gave away a collector's edition Coke bottle. Over five thousand local Freecycle groups around the world provide this service to over 7.5 million members. Group moderators work for free.

My wife and I use an online service called Chip Drop. It matches arborists who generate trucks full of wood chips with people who use wood chips for garden mulch. This business-savvy form of generosity saves arborists fuel and dump fees. Chip Drop now operates in more than twenty US cities.[11] Our electric utility also maintains a waiting list for the wood chips resulting from their seasonal tree trimming.

Craigslist also has a free exchange dimension.[12] Type "free" into the regional site search box and see what people are offering. Also, check out their rideshare page. The 2012 movie *Craigslist Joe* shows how Joseph Garner lived off Craigslist for a month.

And for bibliophiles, BookCrossing offers a way to gift your wonderful books so you can track their journey.[13]

DONATING TO ENVIRONMENTAL CAUSES

Any discussion of green money should include donations to proactive environmental organizations. GuideStar, a nonprofit that tracks revenues, assets, and transparency of other nonprofits, reports that there are over twenty thousand conservation and environmental education nonprofits with a US presence.[14] This book is full of worthy opportunities to help create collective Handprints through nonprofit environmental organizations.

Friends of the Earth (FOE) is comprised of seventy-five member groups, including Center for Environment in Bosnia-Herzegovina and the Russian Social Ecological Union. FOE participates in United Nations climate change negotiations, advocating support for the vulnerable people of the world as we phase out carbon emissions. Projects include stopping oil pipelines, stopping plastic waste dumping in the Global South, advancing vegetarian school meals, convincing grocery chains and their suppliers to avoid pollinator-toxic pesticides, and protecting the world's largest sockeye salmon run in Bristol Bay, Alaska.[15] My contribution to FOE creates a Handprint and reduces a Footprint at the same time. Individuals donated about 20 percent of FOE's 2019 income of $3.1 million (2.8 million euros).[16] That individual donation percentage is relevant because foundations require the recipient to raise matching funds.

Greenpeace has a colorful track record of "bearing witness" to environmentally dangerous and degrading activities. Its ship the *Arctic Sunrise* cruised the US Atlantic coast, bringing attention to offshore oil drilling. Trader Joe's, with Greenpeace encouragement, has reduced its plastic packaging. The First Nations of Canada allied with Greenpeace to defeat the Energy East pipeline from the Alberta tar sands to Eastern Canada. The International Transport Workers Association partnered with Greenpeace to gain better labor practices while curbing illegal tuna fishing and overfishing. Greenpeace receives nearly all its income from some 2.9 million members.[17] No wonder Greenpeace fund raisers ask us to pledge a monthly donation.

The Pachamama Alliance has a two-part mission. Founders Bill and Lynne Twist and John Perkins committed to partner with indigenous people of the Amazon rainforest in responding to threats to their lands and culture from oil and other extractive industries. The indigenous elders in turn requested allies in the Northern Hemisphere who would "change the dream of the modern world" from overconsumption to a culture that honors and sustains life.[18] I experienced the Alliance's eight-week online interactive Game Changer Initiative as a boot camp for peaceful warriors. The Alliance's two-hour Introduction to Drawdown seminar engages participants in Paul Hawken's collection of strategies to curb carbon flow into the atmosphere (see appendix 1).

Myriad national and local nonprofits create Handprints with our dollars, our volunteer hours, and our awareness.

PHILANTHROPY

Thousands of parks are gifts from those who would enhance the commons. George Dorr gave his land to form Acadia National Park, and John Rockefeller gave of his land in Wyoming for Grand Teton National Park.[19] Let us keep that kind of option on the table, perhaps as a land trust in our wills. The commons may be more in need than our grown children.

Gregory Carr, tech-boom millionaire cum philanthropist cum committed angel, took the Gorongosa National Park on as a project. Starting in 2005, working with both the people and the government of Mozambique, Carr committed up to $40 million over thirty years to resuscitate the park. Local elders were respectfully approached, villagers employed, tree nurseries established, and large animals reintroduced to their historic range. From the outset, Carr and the Mozambique government shared the management—but not ownership—of the park. This is part of Mozambique's strategy for ecotourism: to help fuel a sustainable, nonextractive economy.[20]

Panthera, with its founder Thomas Kaplan, initiated the Global Alliance for Wild Cats in 2014. Panthera employs one hundred people in forty-seven countries to protect tigers, lions, jaguars, cheetahs, pumas, leopards, and snow leopards. Commitments from four philanthropists from China, India, the United Arab Emirates, and the United States thus far total $80 million over ten years.[21] Their "community-run conservancies" to protect the snow leopard in Tajikistan and Kyrgyzstan forge alliances with local hunters, where the central government is unable to protect the cats.[22]

In 2009, Stanford University announced that Tom Steyer and his wife, Kat Taylor, endowed the TomKat Center for Sustainable Energy with $40 million. Rather than buying a building, this gift helps professors and students move quickly from science to real-world application. Projects include demonstrating a full-scale efficient wastewater-to-drinking-water treatment plant, field testing e-bikes and e-scooters, and commercializing a low-drag system for long-haul trucking.[23]

While wealth has the potential to do good for the earth, Naomi Klein, in *This Changes Everything*, cautions us that green philanthropy may be skewed by the capitalism that created the wealth in the first place.[24] Other kinds of Handprints are likely needed to change the system.

HANDPRINT OPPORTUNITIES

4.1 Set up an automatic monthly giving plan to foster your favorite environmental organization's mission and stability. What flavor do you love?

4.2 Volunteer with a local food bank. Find it through the Feeding America website.[25]

4.3 Donate reusables to a reselling outlet. Think Value Village, Goodwill, or St. Vincent de Paul.

4.4 Share your garden and orchard surplus with neighbors, the homeless, or Urban Gleaners.

4.5 Use Freecycle or Craigslist to give and receive useful things. Keep useful things out of the landfill.

4.6 Participate in a skill exchange, sometimes called a time bank. Check out the Toe River Skill Exchange in Burnsville, North Carolina, and Time Bank Santa Cruz in California. Where is the generosity? I give you dignity and you give me respect when our time is valued equally. If your community lacks a skill exchange, start one.

4.7 Do a giveaway to mark any special occasion—a graduation, a new job, retirement, a wedding, a clean bill of health after a major illness. Target specific gifts to those who need them. Make special gifts for those to whom you are especially thankful.

4.8 Bequest some of your assets to an environmental cause in your will. Do your children and siblings really need it all?

4.9 Donate land for a park, create a foundation, or offer a prize for a breakthrough accomplishment.

IN YOUR JOURNAL

How will *your* generosity create Handprints?

5

INVEST IN SUSTAINABILITY

We invest a lot of money in ourselves and our institutions. Our retirement funds. Our endowments for churches and colleges. Our mutual funds. Our stock market investments. This money finances every form of industry and civic project. Wouldn't it be good if we could use our collective ownership to serve the health of our planet?

We can.

SOCIALLY RESPONSIBLE INVESTING

At a solar conference in the early 1980s, I heard Hazel Henderson, the keynote speaker, explain the distinction between monetized and unmonetized aspects of our society. We monetize a day's labor for General Motors; we do not monetize the same day of a mother's care for her child. Henderson went on to explain how we could choose to avoid investing in companies that negatively impact the environment. Afterward, I shifted my mutual fund investments to the Calvert Social Investment Fund. For a decade or so, my returns were on par with other modest-risk funds.

More than a fourth of all US investments under professional management now use socially responsible investing (SRI) screens, more than double since 2012, $12 trillion in 2017. That's a "t" as in *Tyrannosaurus rex*.[1]

Divestiture, also known as "negative screening," minimizes our investment Footprint.[2] In the eighteenth century, the Methodists and Quakers divested in slavery, long before the US Civil War. Friends Fiduciary started incorporating Quaker values into its investments starting in 1898. In 1928, the Pioneer mutual fund was launched to screen out companies whose primary business was alcohol or tobacco. In the 1960s and early

1970s, anti–Vietnam War activists promoted divesting military contractors. The number of Dow Chemical Company's investors dropped from ninety-five thousand to ninety thousand in one year, largely due to concerns about napalm.[3]

In the 1980s and early 1990s, ending apartheid in South Africa became an SRI objective. Strategies included shareholder divestment of companies doing business in South Africa *and* shareholder activism motivating banks to stop giving loans (disinvest) to South Africa. These efforts helped motivate the de Klerk government to transition toward a truly representative democracy.

Divestiture was also used as a point of leverage in addressing genocide in South Sudan. The Sudan Accountability and Divestiture Act of 2007 elevated divestiture to the level of government policy. Thirty-five states passed related legislation. In response to the divestment campaign, a dozen companies changed their business practices or left Sudan.[4] South Sudan became an independent nation in 2011.

Positive screening, also a form of SRI, invests in companies that offer sustainable products and services, such as renewable energy. Back in the 1980s, I doubled my $1,000 investment in a solar startup company (which has since been absorbed by others). If you're an investor who is willing to accept risk, your dollars could provide capital for green entrepreneurs. On a large scale, the Clean Trillion initiative, promoted by CERES (formerly Coalition for Environmentally Responsible Economies, based in Boston), will focus $1 trillion per year in US investments on clean energy—potentially a game changer.[5] The target date to reach that level of investment is 2030. This is truly Handprint thinking.

DIVESTING FOSSIL FUELS

In 2012, environmentalist Bill McKibben toured the United States with a bold message: *divest from companies that profit from mining, drilling, and distributing fossil fuels.*[6] Stuart Braman, a veteran of the financial services industry, and his wife heard that message. Two months later, Braman founded Fossil Free Indices (FFI) to track information about companies that profited from fossil fuels, companies to avoid.

A valuable reference produced by FFI is the Carbon Underground 200.[7] This list of one hundred coal companies and one hundred oil and gas companies, ranked by their underground reserves, provides investment portfolio managers a de facto warning label. These companies have the potential to release devastating amounts of carbon into the atmosphere. And these companies face significant loss of valuation from stranded assets as the world becomes more proactive about reducing carbon emissions. That translates into investment risk.

College alumni and church members started lobbying their institutions accordingly. Over 1,100 organizations, representing over $9.4 trillion in assets, have committed to divesting from fossil fuels. These include the likes of Adelaide Bank (Australia); the California Public Employees Retirement System; the Church of England; the Danish Pension Fund; the city of Providence, Rhode Island; the Rockefeller Family Fund; and Stanford University.[8]

Now we can avoid supporting ExxonMobil, AngloAmerican, and Adani. Unfortunately, some large coal, oil, and gas reserves are beyond the purview of most investors—like Gazprom and Rosneft in Russia, China Shenhua Energy, PetroChina, and Coal India.[9]

However, Handprints cascade. The Carbon Underground 200 inspired As You Sow and Corporate Knights to compile the Clean 200 list of publicly traded companies. According to the As You Sow website, "The Clean 200 ranks the largest publicly listed companies by their total clean energy revenues, with a few additional screens to help ensure the companies are indeed building the infrastructure and services needed for [a transition away from fossil fuels] in a just and equitable way."[10] Yes, the Clean 200 includes wind energy companies like Vestas but also information technology companies like Itron and Lenovo (maker of my laptop), electric car companies like Tesla, and real estate companies like Prologis, which "provides industry-leading energy-efficient buildings" for corporate clients.[11]

SHAREHOLDER ACTIVISM

The third form of socially responsible investing (after negative and positive screening) is *shareholder activism.* A maturing cadre of investors and their representatives now champion social and environmental actions at company meetings, often with a financial rationale. For example, shareholder activists might propose that the corporate fleet of cars be transitioned from gasoline to electric fuel to avoid the risk that a price may be placed on carbon through cap-and-trade laws (chapter 19).

In 2002, CERES first made the case that climate change is relevant to the fiduciary responsibility of a company or nation. The flagship project of CERES is the Investor Network on Climate Risk (INCR), which now includes 130 institutional investors. Assets controlled by the network increased from $13 trillion in 2015 to $17 trillion in 2017.[12] Adding to that clout, in 2012, the INCR joined with regional investor groups in Europe, Australia/New Zealand, and Asia to form the Global Investor Coalition on Climate Change. This is not tag football anymore!

The Interfaith Center on Corporate Responsibility (ICCR) also brings to bear the investments of its members. Over three hundred faith- and value-based member organizations collectively represent over *$200 billon* in invested capital. The ICCR includes Catholic churches, Muslim trusts, unions, pension funds, and college endowment funds.[13] ICCR facilitates dialogues as well as shareholder resolutions. In 2015, ICCR engaged

- eighty-one companies regarding reducing greenhouse gases, deforestation, and climate risk;
- twenty companies relating to hydro-fracking;
- eleven companies relating to sustainable palm oil;
- six companies relating to packaging and recycling;
- twelve companies relating to toxins and chemical safety; and
- nineteen companies relating to corporate water impacts.

In 2018, ICCR launched the Investor Alliance for Human Rights with over $3.5 trillion in combined assets, all aligned with the United Nations Guiding Principles of Business and Human Rights. ICCR's goal, as it approaches its fiftieth anniversary in 2021, is to "mainstream" its work into the investment community.[14]

As You Sow specializes in crafting socially and environmentally responsible shareholder resolutions to support CERES, INCR, ICCR, and other investors. An intervention may have several outcomes. A vote of the shareholders raises awareness of an issue. If a sufficient percentage of votes are in favor of a resolution, it may be introduced at the following year's meeting. The company can negotiate with activists, who may withdraw a resolution—possibly a best-case outcome. On occasion, the company may foster change in the whole industry. Rarely, an activist resolution will pass.

What can shareholder activism actually accomplish? Layne Christensen, a water infrastructure management company, agreed to issue annual sustainability reports. Simply reporting sustainability progress, a frequent shareholder proposal, empowers staff to attend to sustainability issues. A 2011 *GreenBiz* article reported that four companies, when confronted by shareholders, withdrew pending plans to use fracking, a strategy to extract oil and gas that endangers groundwater.[15] As You Sow's reports and resolutions are motivating Chevron, Exxon, Kinder Morgan, and Range Resources to reduce climate-changing methane leaks.[16] In 2014, Norwegian Statoil dropped plans to develop its Connor project in Alberta's tar sands, putting a significant pause on removing forty thousand barrels of oil a day from the earth.[17] A proposal to report on financial risks of using coal to generate electricity commands attention in the boardroom. With the urging of shareholders, FirstEnergy committed to retire nine coal-fired power plants in West Virginia, Pennsylvania, Ohio, and Maryland.[18]

In the consumer products arena, the consumer electronics industry— including Acer, Apple, and Best Buy—now cooperates to take back and recycle old computers. In 2013, McDonalds and Dunkin' Brands agreed to phase out polystyrene cups. In 2014, Colgate and Procter & Gamble agreed to make most of their packaging recyclable. In 2017, Unilever agreed to make all its consumer product packaging recyclable; Target committed to phase out polystyrene foam from its home-delivery packaging; and KFC, Burger King, and Wendy's agreed to only purchase chicken raised without medically important antibiotics (as specified by the Food and Drug Administration).[19]

WHERE BLACKROCK GOES

Lest all these SRI investment strategies seem like they are throwing glasses of water on a burning house, take note. In a letter written in January 2020 to the CEOs of the companies BlackRock Inc. invests in, Larry Fink, chairperson and chief executive officer, conveyed this message: "In the discussions BlackRock has with clients around the world, more and more of them are looking to reallocate their capital into sustainable strategies. If ten percent of global investors do so—or even five percent—we will witness massive capital shifts."[20]

BlackRock Inc. is the *world's largest* investment manager of endowments and pension funds, controlling over $6 trillion in assets.[21] "Given the groundwork we have already laid engaging on disclosure, and the growing investment risks surrounding sustainability, we will be increasingly disposed to vote against management and board directors when companies are not making sufficient progress on sustainability-related disclosures and the business practices and plans underlying them." In 2019, BlackRock withheld support or voted against 4,800 directors at 2,700 companies. This is a message that sends not just ripples but tidal waves through the boardrooms of the corporate world.

HANDPRINT OPPORTUNITIES

5.1 Invest with an environmental or social justice screen. The US SIF (formerly the Social Investment Forum) offers a place to start. Their online member directory lists investment management members such as Trillium and Earth Equity Advisors. Socially responsible investment mutual funds include Domini, Pax World, and the nonprofit Green Century Funds.[22]

5.2 Make it personal. One woman I know worked up the courage to talk with her elderly father about his wealth. She was the only one who could show him that it mattered how he invested his millions. That became another divestment story.

5.3 Invest directly in earth-friendly products and services. Energy efficiency, renewable energy, organics, equipment rental companies . . .

5.4 Invest in Clean 200 companies.

5.5 Divest from Carbon Underground 200 companies. Avoid supporting companies that extract carbon from the earth's crust.

5.6 Insist that your alma mater, retirement fund, and faith community align with socially responsible investment strategies of CERES, As You Sow, and the Interfaith Center for Corporate Responsibility.

IN YOUR JOURNAL

How will *you* help create Handprints with your investments?

6

WORK FOR
ENVIRONMENTAL JUSTICE

What if we as a world could agree on one document that represents humankind's best intentions? What if it drew on the collective cultural wisdom of Earth's people? What if it embodied the principles of environmental justice? Such a document exists.

PREAMBLE TO THE EARTH CHARTER

We stand at a critical moment in Earth's history, a time when humanity must choose its future. As the world becomes increasingly interdependent and fragile, the future at once holds great peril and great promise. To move forward we must recognize that in the midst of a magnificent diversity of cultures and life forms we are one human family and one Earth community with a common destiny. We must join together to bring forth a sustainable global society founded on respect for nature, universal human rights, economic justice, and a culture of peace. Towards this end, it is imperative that we, the peoples of Earth, declare our responsibility to one another, to the greater community of life, and to future generations.

Source: "The Earth Charter," Earth Charter Commission, accessed May 6, 2020, https://earthcharter.org/wp-content/uploads/2020/03/echarter_english.pdf?x28510.

THE EARTH CHARTER

The Earth Charter has its roots in a 1987 call by the World Commission on Environment and Development to set "new norms" to guide sustainable development.[1] The Baha'i community, supporting the initiative, stated, "Only discourse at the level of principle has the power to invoke a moral commitment, which will, in turn, make possible the discovery of enduring solutions to the many challenges confronting a rapidly integrating human society. There are spiritual principles, or what some call human values, by which solutions can be found for every social problem."[2] While the Earth Charter was discussed at the 1992 Earth Summit in Rio de Janeiro, its constituency included indigenous nations and nongovernmental organizations that were largely ignored by diplomats. It was not directly tied to economic development and contained no system for implementation or enforcement. This guide to responsible action in civil society, business, and government was formally approved after a long consensus process on June 29, 2000.[3]

The first of sixteen principles reads, "Respect Earth and life in all its diversity. a. Recognize that all beings are interdependent and every form of life has value regardless of its worth to human beings. b. Affirm faith in the inherent dignity of all human beings and in the intellectual, artistic, ethical, and spiritual potential of humanity."[4]

The UN-mandated University of Peace in Costa Rica hosts Earth Charter International headquarters. Around the world, over a hundred partners and affiliates collaborate on projects that align with the Earth Charter, including the Arab Foundation for Environment and Development and the Portuguese Association for Environmental Education.[5] The Earth Day Network "provides civic engagement opportunities at the local, state, national and global levels." Green Cross International, a legacy of Mikhail Gorbachev, seeks to build "political will for the deep-rooted societal change that is capable of containing and then reversing the effects of climate change and environmental degradation." Amana-Key, based in Brazil, focuses on the "conscious evolution of leaders and management of governmental and non-governmental organizations." The Forum on Religion and Ecology at Yale University provides a portal for "religious and spiritual communities on ecological issues."

Environmental justice is one dimension of the Earth Charter's scope, especially as reflected in the charter's seventh principle: "Adopt patterns of production, consumption, and reproduction that safeguard Earth's regenerative capacities, human rights, and community well-being."[6] Issues include water quality, landfills, oil pipelines, and breaches of treaty rights.

Each injustice presents Handprint opportunities through awareness, activism, lobbying, research, and forging healthy ways forward.

FLINT'S WATER

The water crisis in Flint, Michigan, presents a modern environmental justice case study. The city of Flint was already experiencing financial problems when the unelected city manager decided to save money by switching the water supply from the Detroit municipal system to the Flint River. Complaints about smell and turbidity began immediately. It took a little longer to measure the toxic lead levels in Flint's children.[7] By the way, the population of Flint was 58 percent African American in 2014 when the crisis began.[8]

Who do we have to thank for turning this crisis around? LeeAnne Walters, a mother who saw weird rashes developing on her children, contacted the Environmental Protection Agency. Miguel del Toral, the agency's Midwest manager, responded to Walters's call, confirming the problem and bringing in state agencies. Marc Edwards of Virginia Tech (with the aid of students) paid $150,000 of his own money to carefully test for lead in the water around town. Dr. Mona Hanna-Attisha, head of the pediatric residency program at Hurley Medical Center in Flint, compared the lead levels in children's blood before and after the switch.[9] Documentarian Michael Moore informed a wider world about the official cover-up of the problem in his movie *Fahrenheit 11/9*. And one more person we may yet want to thank: Jonathon Lubrano published a potential strategy for paying for the needed infrastructure upgrades. His 2017 report in the *William and Mary Environmental Policy Review* is titled, "Water, Lead and Environmental Justice: Easing the Flint Water Crisis with a Public Water Contamination Liability Fund."

I'd happily give each of these folks an environmental justice Handprint award. Protests and enquiries change our awareness of an issue. That echoes forward.

LANDFILLS

In 1948, New York City needed a place to put its garbage. It chose Fresh Kills, a 2,200-acre rural area on the west side of Staten Island, the poorest borough of the city. The landfill's odor fueled hundreds of protests. Still, the governor of New York vetoed legislation to close the landfill. Groups like Staten Island Anti-Garbage Organization and Staten Island Citizens for Clean Air formed. At its peak in 1986–1987, Fresh Kills received 29,000 tons of garbage *per day*. It took until March 22, 2001, to unload that last barge of garbage.[10] Did local protests stop the Fresh Kills landfill? No.

But concerns about municipal solid waste in the United States gave birth to the environmental justice movement. The commissioners of Contra Costa County, California, were confronted by the group We Have Enough Waste when they discovered that all the proposed sites for a municipal waste dump were in the poorer, and ethnically diverse, east side of the county. The Reverend Benjamin Chavis Jr., formerly of the National Association for the Advancement of Colored People, coined the term "environmental racism" to describe the problem. The Citizens Clearinghouse for Hazardous Wastes, by advocating composting, recycling, and better waste management practices, helped stop the creation of dumps from Eagle Mountain, California, to Sparta, Georgia.[11] In the United States, linking the environmental movement to the civil rights movement proved critical in furthering the goals of both movements.

Activism from citizens and groups concerned about landfills ultimately led to environmental legislation. The Solid Waste Disposal Act of 1965 and its direct, and still-relevant, descendant, the Resource Conservation and Recovery Act of 1976, became the United States' primary laws for the disposal and recovery of municipal and hazardous waste. The act set out to regulate solid waste, hazardous waste, and underground storage tanks. Regulations established a "cradle-to-grave" approach to hazardous waste, and the Federal Hazardous and Solid Waste Amendments of 1984 phased out land disposal of hazardous waste altogether.[12]

Over time, our growing disgust with landfills has fueled municipal recycling programs in most cities. In 2015, almost 35 percent of our national municipal solid waste was diverted from landfills.[13]

ENVIRONMENTAL JUSTICE IN INDIAN COUNTRY

American Indian, Alaskan Native, and First Nation people have been battling environmental injustice for centuries. In 1864, the peach orchards in Canyon de Chelly were destroyed by the US Army to starve the Navajo into submission.[14] Buffalo were slaughtered from trains.[15] The Black Hills of South Dakota were taken from the Lakota in the 1870s, largely to gain access to "lumber" and "minerals" (gold).[16] In the 1940s, 1950s, and 1960s,

Navajos were subjected to the health hazards of uranium mining.[17] Six dams on the Upper Missouri River inundated about 550 square miles (142,000 hectares) of Indian land in the 1950s and 1960s.[18] In 1979, the dam at the Church Rock uranium mill on the Navajo reservation collapsed, becoming the nation's largest radioactive waste spill.[19] Dina Gilio-Whitaker, author of *As Long as Grass Grows: The Indigenous Fight for Environmental Justice, from Colonization to Standing Rock*, tells us that this level of environmental injustice—along with a pattern of removal from the land and sources of traditional food, a trail of broken treaties, and a context of colonial conquest—meets the criteria for genocide.[20]

Recognizing this historical reality helps to frame environmental justice Handprints in and around Indian Country. While providing reparations for past wrongdoings may be beyond the means of most of us, we can at least partner with Indians on projects *they* want to work on, supporting *their* sovereignty and leadership. Gilio-Whitaker describes several strategies that go beyond custodial (hierarchical) aid programs, such as land trusts, archeological resource protection, protection of sacred lands, and shared management of parks.

Government-to-government cooperation is clearly part of the environmental justice playbook. The Timbisha Shoshone comanage 7,754 acres with Death Valley National Park. A shared objective: recovery of groves of honey mesquite and piñon pine—traditional sources of food.[21] The Navajo in Arizona share in the management of both Canyon de Chelly National Park and Navajo National Monument.[22] In Alaska, the Indian Environmental General Assistance Program, administered by the US Environmental Protection Agency, helped the Akiak Yup'ik community execute a recycling program, motivated by concerns about leaching from the landfill.[23]

Indians in California established important precedents in working with the California Coastal Commission. The Acjachemen Nation (Juaneño Band of Mission Indians) and a coalition of environmentalists convinced the Coastal Commission to deny a permit for a new highway. The unmitigable impact on a sacred site (an ancient village known as Panhe) proved to be compelling. As a result, San Mateo Creek and the acclaimed Trestles surfing site were protected.[24] A few years later, when developer Newport Banning Ranch proposed building a commercial center on another stretch of beach, a respectful prior consultation with local Tongva people secured protection for local sacred sites—as well as restoration of nearby land that had long been degraded by oil drilling.[25]

In numerous places, native peoples have secured access to, and protection of, land outside reservation boundaries. The Native American Land Conservancy used a land trust to protect the Old Woman Mountains in the Mojave Desert east of Los Angeles. Establishing rights of nature is another concept consistent with traditional Native American values. One initiative seeks to secure personhood for the Colorado River. There is precedent for this: In 2013, in New Zealand, Maori prompting yielded personhood for Whanganui and Te Urewera National Parks. In 2017, India granted both the Ganges and the Yamuna Rivers personhood. Also, when consulted during the framing process, tribes can be protected by community rights ordinances of adjacent counties and towns (chapter 3).[26]

In the realm of activism, the informal Cowboy Indian Alliance blends the interests of ranchers and Indians on the northern Great Plains. This movement successfully fought a plan by Honeywell to test weapons in South Dakota's Black Hills and a plan to build a new railroad line to carry coal in Wyoming. This loose coalition also catalyzed efforts to stop the Keystone XL pipeline that would have stretched from the tar sands in Alberta to Steel City, Nebraska—a movement I joined in the summer in 2013.[27]

STANDING ROCK

The encounter just north of the Standing Rock Reservation in North Dakota offers an example of how environmentalists can support Indians in environmental justice causes. For 314 days, deep into the winter of 2017, "water protectors" blocked the 1,172-mile Dakota Access Pipeline from being built across the Missouri River in North Dakota.

My wife and I paid for the bus ticket of one of the twenty thousand people who brought supplies and shared ceremony with the Standing Rock Lakota Nation. We also raised money for a congregation to orient people from Japan, Germany, Israel, Australia, Serbia, Russia, Samoa, Hawaii, New Zealand, and the rest of the United States who also stood beside "the people."

Standing Rock helps us redefine environmental justice from an indigenous point of view in terms of not only race but also colonization.[28] The gathering along Íŋyaŋwakağapi Wakpá (the River that Makes Sacred Stones) near where it meets the Missouri River was about treating water as sacred, *mni wiconi* (pronounced mni wichoni); about violation of the 1851 Fort Laramie Treaty, one of many treaties the United States has broken;

about respecting buried remains, which were bulldozed; about asserting sovereignty in the wake of genocide; and about forging alliances with indigenous peoples across the planet.

The stand at Standing Rock allied those who would confront an invasion and those who would protect the environment—in that order. In other words, Native Americans took the lead. Common cause with those who live closest to the land raised awareness in faraway cities—and raised the price for degrading the environment. The confrontation ended with the remaining occupiers burning their lodges. But the stand they (we) took empowers new stands elsewhere. Gilio-Whitaker asserts, "Effective partnership with allies in the environmental movement will provide the best defense for the collective wellbeing of the environment and future generations of Americans, Native and non-Native alike."[29]

Such is the nature of effective environmental justice Handprints.

ONE CONGREGATION'S INITIATIVE

Portland, Oregon's, Wy'east Unitarian Universalist Congregation found a way to support environmental justice within the scope of its weekly service. During the 2014–2015 church year, the congregation donated one collection a month to a specific organization, usually local. Invited speakers, in turn, helped congregants better understand the issues.

A host of environmental justice organizations operate mostly outside the awareness of mainstream society. Kyle Curtis of Outgrowing Hunger explained community gardens. Jon Ostar presented several cases in which OPAL (Organizing People, Activating Leaders) Environmental Justice Oregon provided legal help to beleaguered communities. Elevating the voices of people affected by the Portland Harbor Superfund site, the Portland Harbor Community Coalition includes local branches of the American Indian Movement, the Asian Pacific American Network, the Czech School, and the Iraqi Society. Verde is helping transform a twenty-five-acre landfill into a park with the Native American Youth and Family Center, Hacienda Community Development Corporation, and the Cully Association of Neighbors.[30]

Some of the issues are of global importance. Rose Highbear, founder of Wisdom of the Elders, shared stories of Alaskan natives who are facing encroaching oil drilling. Jasmine Zimmer-Stucky, of Columbia Riverkeepers, explained how small communities, needing to preserve their quality of life, stop trains from exporting coal overseas, where it is burned and released into the atmosphere. John Audley of Sustainable Northwest described how rural Oregonians, who once were dependent on clear-cutting forests, group together to obtain Forest Stewardship Council certification to cut costs and access high-return furniture and cabinetry markets.[31]

In one year, the sixty-member Wy'east Congregation gave $4,000 to support these causes.[32] What if thousands of congregations did this kind of thing? Micro Handprints add up to macro Handprints.

HANDPRINT OPPORTUNITIES

6.1 Endorse the Earth Charter, either as an individual or as an organization. This entails an active decision to live and act by its principles. Support Earth Charter International.

6.2 Attend rallies announced by environmental justice organizations in your area. Being heard is critical. Help raise money for them as well.

6.3 Show climate justice films at your school or congregational or neighborhood meeting. Think *Mossville: When Great Trees Fall* about Stacey Ryan's refusal to sell his land for the expansion of a petrochemical complex in Louisiana. Think *Awake: A Dream from Standing Rock*. Check out lists of such eye-opening films online. These stories need to be told and retold.

6.4 Support media that report environmental justice issues. Read *Mother Jones* and *Yes!* magazines and *Censored News*, a long-running blog edited by Brenda Norrell.[33] Watch the *Rachel Maddow Show* on MSNBC. Listen to Thom Hartmann on progressive radio stations and podcasts.

6.5 Call attention to industrial pollution in rivers and groundwater, leaching from mine tailings, leaks from pipelines, contamination of drinking water, derailments involving toxic chemicals, toxic spraying, windblown coal dust, siting and maintenance of landfills, effluents of concentrated animal feeding operations, and action plans for Superfund sites. Such events and activities disproportionately impact the rural and the poor.

6.6 Donate to the Deep South Center for Environmental Justice. This center bridges the gap between Gulf Coast region community experience and academic knowledge, and between community capacity building and responses to events like Hurricane Katrina and the *Deepwater Horizon* oil spill of 2010. Students at the center earn certifications in lead abatement, asbestos removal, mold remediation, and hazardous waste operations and emergency response—with an 85 percent permanent employment placement rate.[34]

6.7 Donate to We Act for Environmental Justice. Based in New York City, WE ACT's agenda includes electric school buses, reducing lead exposure, and banning mercury from skin products.[35]

6.8 Support indigenous environmental justice organizations such as these:

- **Big Elk Native American Center** supports the Omaha (Umonhon) and other Native Americans living off-reservation near Omaha, Nebraska, while asserting legal and treaty rights. The center's Nature Academy will teach young people *from any background* traditional ways to grow medicinal and food plants.[36]

- **First Nations Development Institute** projects include campgrounds to promote ecotourism on the Fort Belknap reservation in Montana and determining the market value of carbon stored by habitat restoration and preservation on the Lower Brule Sioux tribal land in South Dakota.[37]

- **Honor the Earth** creates "awareness and support for Native environmental issues and [develops] needed financial and political resources for the survival of sustainable Native communities." Honor the Earth hosts a legal defense fund for those opposing the Enbridge pipeline in northern Minnesota.[38]

- **Indian Law Resource Center** "provides legal assistance to indigenous peoples," drawing the connection between "indigenous land rights, environmental protection and human rights." The center helped form the Yukon River Inter-Tribal Watershed Council.[39]

- **Movement Rights**, an international indigenous-led organization, "assists communities confronted by harmful corporate projects to assert their right to make important decisions that impact them by passing new laws that place the rights of residents (and nature) above the claimed legal 'rights' of corporations."[40]

- **The Native American College Fund** is the "nation's largest charity supporting Native student access to higher education [providing] scholarships, programming to improve Native American student access to higher education, and the support and tools for them to succeed once they are there."[41]

- **Red Cloud Renewable Energy Center** "is a hub for green job training in Indian Country where tribal members from across the country receive training in a wide spectrum of renewable energy and sustainable building applications from fellow Native Americans."[42]

IN YOUR JOURNAL

How will *you* further the cause of environmental justice?

7

SUPPORT
SUSTAINABLE DEVELOPMENT

News flash: the world agrees on sustainable development. On September 15, 2015, the United Nations Sustainable Development Summit, with 150 nations represented, adopted 17 goals and 169 supporting targets to chart the course of sustainable development through 2030.[1] The parties required over three years of intense work to "cover the whole sustainability agenda: poverty, human development, the environment, and social justice." These life-affirming goals and targets apply to *all* nations. They apply to businesses, education, and communities as well as government. In coming to this accord, we forge a working relationship between the idealistic constituency of "sustainability" and the economic aspirations of "development."[2]

The Sustainable Development Goals (SDGs) are more operational than the Earth Charter (chapter 6) and represent more of a global work plan than a law or treaty. SDG 1 proposes to "end poverty in all its forms everywhere," including (Target 1.4) providing access to microfinance. SDG 4 seeks to "ensure inclusive and equitable quality education and promote lifelong learning opportunities for all," including (Target 4.5) eliminating gender disparity in education, (Target 4.6) providing literacy and numeracy education for all youth, and (Target 4.7) teaching lifestyle and nonviolence skills needed to promote sustainable development. SDG 7 challenges us to "ensure access to affordable, reliable, sustainable and modern energy for all," including (Target 7.2) increasing the share of renewable energy in the global energy mix and (Target 7.3) doubling global energy efficiency. See appendix 2 for a list of all seventeen SDGs.

In lieu of an enforcement mechanism, the SDG agenda has the force of goodwill and practical acumen. For instance, Finland announced it will achieve carbon neutrality by 2035, and Nigeria will engage half a million children in learning the SDGs.[3]

This book offers a cornucopia of ways each of us, in our own way, can help bring the Sustainable Development Goals and targets to fruition.

SUSTAINING ENTREPRENEURS

We can fight poverty (SDG 1) and hunger (SDG 2) and promote decent work environments (SDG 8) by supporting the work of established organizations that use proven effective strategies.

The fair trade movement encompasses many organizations, including Fair Trade USA, Equal Exchange, the World Fair Trade Organization, Cocoa Life, Fair for Life, and Fairtrade International and its counterpart Fairtrade America. All these organizations seek to promote the welfare of workers and small landowners.[4]

The Oromia Coffee Farmers Cooperative Union is located in Ethiopia. Its Fairtrade certification dates to 2002. Members of the union are trained in organic farming and composting of waste. In addition to money paid for workers' produce and labor, a Fairtrade premium provided to the cooperative helps improve the quality of workers' lives. That premium helped build more than fifteen schools, four health clinics, and fifty-six clean water supply stations.

That story, and others like it, thrives behind that little blue-and-green Fairtrade logo on Highground Coffee, Thanksgiving Coffee, and Red Diamond Organic Coffee.[5] Fair trade products range from asparagus to avocados, coffee to cotton, grapes to geranium oil, and soda beverages to olive oil.

Heifer International supports small-scale farming (SDG 2). Among their programs, Heifer helped coffee growers in sixteen countries in Latin America. Family farmers typically suffered lean months due to cycles in the harvest, variability of commodity prices, weather, and crop disease. Strategies included teaching product improvement skills, providing vegetable seeds, diversifying the salable food commodities with bees and livestock, and providing access to microcredit.[6]

Microcredit has a platform in the developed world. In 2004, while on a sabbatical from PayPal, Premal Shah visited India and decided to use his Silicon Valley experience and the principles of microfinance to start Kiva.[7] Kiva partners with local organizations around the world to provide individual loans ranging from $400 to $1,000. For instance, UpEnergy distributes efficient stoves, water purification systems, and solar lights in Uganda, Rwanda, and Central Africa. Kiva loans help shops (borrowers) purchase inventory. Each loan is provided to the borrower without a fee.[8]

In 2015, I made my first microloan through Kiva to Hasan, who lives on the West Bank, via Palestine for Credit and Development. Hasan put a passive solar water heating system on his roof and bought a water purification system. My $25 was part of a $1,500 loan. In this case, the money is "predisbursed," so Hasan did not need to wait for us investors to find his story on the website to move ahead on his project. Palestine for Credit and Development accepts the risk if the money is not raised within ninety days or if Hasan fails to repay the loan.[9] While the banks would consider such a loan "high risk," the repayment record for such loans is well over 90 percent.

Twelve months after making my Kiva loan, I had a choice: reclaim my $25, donate it to Kiva to cover overhead costs, or initiate another loan. My next loan, in 2017, helped fund a $4,000 college education for a young woman in Kenya, who now has a ten-year period for repayment. This Handprint supports SDG 5, gender equality.

EDUCATION FOR SUSTAINABLE DEVELOPMENT

Twenty-six percent of the world's population is fourteen years old or younger.[10] Thus, education (SDG 4) is fundamental to truly sustainable development. The UN Decade of Education for Sustainable Development (ESD) ran from 2005 to 2014.[11] Japan, Denmark, and Sweden provided special funding for this international cooperative effort.

Types of training encompassed by ESD include teacher training, system thinking, local problem solving, investigations and discovery, critical thinking, and working with multiple stakeholders and disciplines.[12] The *Education for Sustainable Development Toolkit* by Rosalyn McKeown, prepared as part of this decade-long initiative, is informally referred to as a bible for sustainable education. She concludes her tool kit by recommending the following:

- ESD must be locally and culturally relevant, reflecting local conditions.
- ESD must be created through a public process, allowing for community visions of sustainability.

- Communities and schools should work together to attain sustainable goals.[13]

ESD now enjoys widespread international acceptance. Mauritius and Costa Rica have integrated it into national sustainable development policies.[14] ESD is part of Bhutan's gross national happiness initiative (chapter 18). Kenya included ESD in both its five-year (2013–2018) education revitalization plan and its 2030 national sustainability plan.[15] Environmental education has been enacted into law in Brazil, Cambodia, India, Japan, Kazakhstan, Scotland, and Sweden.[16] Thousands of local governments have embraced ESD in one way or another—Hamburg, Germany, and Barcelona, Spain, being notable examples.[17] In Albania, 2,500 elementary school teachers and 1,000 principals have been trained in sustainable development.

Education for sustainable development is the guiding theme for Regional Centers of Excellence established around the world under the auspices of the UN Institute for the Advanced Study of Sustainability in Tokyo. Most are hosted by a local college. My local Greater Portland Sustainable Education Network welcomes scores of sustainability-inspired schools and nonprofits, convening dozens of events in a constant churn of networking and creativity. Other US Regional Centers of Excellence have been established in Georgetown County, South Carolina (Coastal Carolina University); Grand Rapids, Michigan; Greater Atlanta, Georgia; Greater Burlington, Vermont; North Texas (University of Texas at Arlington); and Shenandoah Valley, Virginia. When I saw a map with orange dots representing more than 170 Regional Centers of Excellence, it hit me viscerally that the whole world is in the process of becoming more enlightened about sustainability.[18]

EDUCATING GIRLS

SDG 4 (education) and SDG 5 (gender equality) both highlight the need to educate girls, especially where they are traditionally married at a young age.

The Central Asia Institute (CAI), founded in 1996, builds schools "to promote education and livelihood skills, especially for girls and women, in the remote regions of Afghanistan, Pakistan, and Tajikistan." The CAI operates in a part of the world where girls are traditionally kept at home until they are married. In Afghanistan, 35 percent of girls are married before the age of eighteen, 9 percent before the age of fifteen.[19] Now that CAI has over 450 projects in Afghanistan, Pakistan, and Tajikistan, there is a waiting list of five projects for every project completed.[20]

I admit to being charmed by the bright eyes looking out of every issue of CAI's annual *Journey of Hope* magazine. That and the tagline "Educate a girl. Change the world" have motivated me to be a regular donor for at least a decade.

Using a different strategy in East Africa (Kenya, Tanzania, and Uganda), the nonprofit Growth Through Learning gives scholarships to three hundred or more girls to attend secondary school each year.[21] How can we multiply that number by ten thousand?

More than 573 million women and girls are illiterate—over half a billion. The direct benefits of educating young women are compelling in their own right. A woman's income increases 10 to 20 percent with each year of schooling she receives. Children of literate mothers are 50 percent more likely to survive past age five. Children of educated mothers are 50 percent more likely to go to school.[22]

WHY EDUCATION RESULTS IN FEWER CHILDREN—FOUR THEORIES

- **Economics.** Women who have more opportunities to make money have a higher opportunity cost associated with having children.
- **Household bargaining.** Women who can make money outside the home have more bargaining power in the home to determine the size of the family.
- **Ideation.** Women with education are exposed to a world of thinking that allows them to make choices.
- **Health.** Educated women know more about prenatal care and childcare, which leads to a greater confidence that the children they have will survive into adulthood.

Source: Elina Pradhan, "Female Education and Childbearing," *Investing in Health*, World Bank, November 24, 2015, http://blogs.worldbank.org/health/female-education-and-childbearing-closer-look-data.

The statistical links between education and birth rate are compelling. In Ethiopia, girls with eight years of schooling will have, on average, half the number of children as those with no schooling at all. In Ghana, on average, a woman with a high school education will have two or three children—her total fertility rate (TFR). If she has no education, that number would be about six.[23]

Data from Asia and Latin America confirm these trends. In the first decade of the millennium, uneducated women in the Philippines had a TFR of 5.3 children. A high school education brings that TFR down to 3.1 children. In Bangladesh, the TFR moves from 3.6 to 2.5 with a secondary education.[24] In Honduras, a high school education moves the TFR from 4.9 to 2.2.[25] The percentage of young women between fifteen and nineteen years of age who are mothers consistently drops with the number of years of schooling. In Peru, teenagers are half as likely to have children if they stay in school. In Brazil, 40 percent of uneducated, indigenous Brazilian late-teenager women are mothers. That percentage drops to 16 percent with seven to twelve years of school.[26]

When girls and women are educated, families are educated. When families are educated, nations are educated. When nations are educated, their standard of living goes up and their population stabilizes.[27]

CONTRACEPTION

Contraception is a more direct approach to counteracting the stress of unplanned children on local and international ecosystems. The good news is that we have come a long way in being able to control our natural urge to procreate. More than five billion condoms are sold every year.[28] Alan Weisman reports exemplary programs to provide reproductive services in Bangladesh and Thailand. Birth rates are now so low in Japan, Finland, and Italy that policy makers worry about who will take care of the aging populations.

However, in 2018, the World Health Organization reported, "214 million women of reproductive age in developing countries who want to avoid pregnancy are not using a modern contraceptive method." More specifically, "in Africa, 24.2 percent of women of reproductive age have an unmet need for modern contraception." The World Health Organization also reports over 10 percent still need access to contraception in Asia, Latin America, and the Caribbean.

Target 5.6 of the gender equality SDG asserts the need for universal access to sexual and reproductive health care. Barriers include limited access to birth control and health care, especially among the young, unmarried, and poorer segments of the population. The UN Population Fund (also known as United Nations Family Planning Agency) seeks to remove these barriers with access to modern contraceptives, training of health workers, and advocacy to end child marriage. Over 40 percent of the fund's budget comes from private donations.[29]

In his book *Drawdown*, Paul Hawken, referring to his team's analysis of ways to reverse carbon buildup in the atmosphere, concludes, "The number one solution to global warming is empowering girls and women."[30]

HANDPRINT OPPORTUNITIES

7.1 Memorize the seventeen Sustainable Development Goals. Think of them like the Pledge of Allegiance. Or a prayer. Occasions will arise for you to apply them wherever you go.

7.2 Visit Ten Thousand Villages, a nonprofit enterprise founded by Edna Ruth Byler. They follow fair trade principles. Find handmade items from some twenty thousand makers at three hundred stores and online. Or perhaps you are more inclined to volunteer.[31]

7.3 Buy fair trade food, clothing, and other items. Check out Fair Trade USA for products and partners.[32]

7.4 Buy from Native American–owned businesses such as these:

- **Bedré Fine Chocolate** is owned by the Chickasaw nation in Oklahoma. Check out their chocolate bars, premium sauce, caramel pecan sensations, and gift box.[33]
- **Beyond Buckskin** offers original work online from "first artists and original designers." Check out their birch circle earrings and buffalo hide woven moccasins. Based on the Turtle Mountain Chippewa reservation in North Dakota.[34]
- **Eighth Generation** offers wool blankets designed by Native artists. The brand is owned by the Snoqualmie Tribe. The flagship retail store is located in Seattle's Pike Place Market.[35]
- **Native Harvest** offers Minnesota wild rice, maple syrup, and more. This perennial is harvested by hand and grows on water. No soil is disturbed. Proceeds support the White Earth [Indian Reservation] Land Recovery Project.[36]

7.5 Donate to Heifer International. Projects include Fermented Food for Life in Uganda; Improving Income, Nutrition through

Community Involvement in Cambodia; and Rise Up: Building Resilience of Farming Families in the Philippines.[37]

7.6 Make a microloan via Kiva. Check regularly to find a person or project that resonates with you. Once the needed money is raised from a loan provider, another opportunity takes its place online.

7.7 Support organizations that educate girls and women. Check out Camfed, Central Asia Institute, CARE, Educate Girls, Girl Scouts, Forum for African Women Educationalists, Growth Through Learning, and She's the First.

7.8 Support organizations that support family planning internationally. Consider the UN Population Fund and Planned Parenthood Global.

IN YOUR JOURNAL

What steps will *you* take to create sustainable development Handprints?

III

TEACHING AND
LEARNING HANDPRINTS

8

TEACH ECOLOGY

Helping a child—of any age—understand and care for our planet is a wise investment for the future. Learning about nature and sustainability gives us the inclination and tools to make a difference. Educators and outdoors people across the United States are beginning to restore environmental literacy to its rightful place in the upbringing of our citizenry. Camps, schools, zoos, and even ecotourism offer paths toward caring stewardship. Rex Stout once wrote, "Any spoke will lead an ant to the hub."

ENVIRONMENTAL LITERACY

For millennia, indigenous peoples have conveyed environmental wisdom through the oral tradition. For the Hopi people of northeastern Arizona, environmental education is conveyed through their creation story. Sitting in underground ceremonial spaces called kivas, people of all ages learn how the creator god Sotuknang deals harshly with humans who disregard nature. Such stories helped the Hopi persevere in a severe desert environment.

HOPI PROPHECY

In the Hopi creation story, human beings give Sotuknang serious cause for concern. In the first world, people separate themselves from creation, losing their ability to communicate with other animals. In the second world, people lose their relationship with the creator. In the third world, humans start warring on each other. Sotuknang ends each world with a scourging—first by fire, then by earthquake, and then by flood. But there are beings on earth who help people who are willing and able to be in right relationship with creation. In one case, ants kept the people safe underground and taught them about storing food and living in community.[a]

According to the story, we now live in the fourth world.

For fifty years, Thomas Banyacya, a Hopi elder, shared a prophecy associated with the creation story with whoever would listen. Banyacya warned of yet another "purification" if we fail to follow peaceful ways and respect nature.[b] He took his message to the United Nations in 1992. He died in 1999.

---

[a] Bruce Railsback, ed., *Creation Stories from Around the World*, 4th ed. (Athens: University of Georgia, Department of Geology, July 2000), chapter 1 "The Four Creations," http://www.gly.uga.edu/railsback/CS/CSFourCreations.html. (Story collected from Hopi elders by Oswald White Bear Fredricks and Naomi Fredricks in the 1950s.)

[b] Robert McG. Thomas Jr., "Thomas Banyacya, 89, Teller of Hopi Prophesy to the World," *New York Times*, February 15, 1999, http://www.nytimes.com/1999/02/15/us/thomas-banyacya-89-teller-of-hopi-prophecy-to-world.html.

William B. Stapp is known as the father of (modern) environmental education. He was nominated for the Nobel Prize in 1993. As a teacher in Ann Arbor, Michigan, Stapp had a passion for getting schoolchildren out into the environment and a capacity to influence school boards. In 1959, Ann Arbor (Michigan) Public Schools adopted the first comprehensive K–12 conservation and outdoor education program in the United States. It is still going strong today.[1]

The North American Association for Environmental Education offers its *K–12 Environmental Education: Guidelines for Excellence* to educators everywhere. It provides a road map to environmental literacy with expectations for fourth, eighth, and twelfth grades.[2]

Students in kindergarten to grade four learn to ask questions that help them conduct simple investigations and learn about their environment. At the Central Park School for Children in Durham, North Carolina, teachers use the project-based learning model. For example, "students read about storms; write about them; use math and science skills to analyze them; use art and movement to explain them. They'll invite in experts, and go to places where they can learn more about storms."[3]

Students in grades five through eight collect information about their environment in field studies. They learn to think like scientists. In Bakersfield, California, students demonstrated the advantages of using a mix of fresh water and gray water (from showering and washing dishes and clothes) to keep lawns green while meeting state water conservation targets.

Students in grades nine to twelve learn to describe how human sustainability depends on earth systems. Marsh-Billings-Rockefeller National Historic Park invites instructors and students at Woodstock Union High School to conduct projects to "interpret the history and evolution of conservation stewardship in America." One class developed an outdoor classroom on a donated parcel of land.

The state of Pennsylvania integrates environment and ecology into its Standards Aligned System, which is designed to prepare teachers and students for annual assessments and Keystone Exams prior to high school graduation.[4] A high school student, taking a Keystone Exam before graduation, might be asked to "describe how ecosystems change in response to natural and human disturbances (e.g., climate change, introduction of nonnative species, pollution, fires)."[5]

In 2010, Wisconsin developed the Green and Healthy Schools program to implement its environmental education requirements. This program has been emulated in California, Oregon, Maryland, Massachusetts, New Jersey, Tennessee, and Kentucky.[6]

The Oregon nonprofit Ecology in Classrooms and Outdoors deploys "scientists in residence" to support teachers. Sometimes paid for by the local parent-teacher association, this program fills gaps for students with special needs and districts with limited capacity.[7]

ZOOS AND AQUARIUMS

Some zoos and aquariums fill a vital role in our environmental education. In an increasingly urban world, it is hard to love that which we do not know. Of the ten thousand zoos and aquariums worldwide, 239 (in 2019) meet the standards of the Association of Zoos and Aquariums, which requires support of wildlife conservation.[8]

The Toledo Zoo in Ohio is engaged in the Mariana Avifauna Conservation program. Invasive egg-eating brown tree snakes have devastated native bird populations in the Pacific Mariana Islands. While relocation efforts continue for the threatened species, the Toledo Zoo hosts a reserve population of the white-throated ground dove, the golden white-eye dove, the bridled white-eye dove, and the Mariana fruit dove.[9]

The Bronx Zoo, Central Park Zoo, Prospect Park Zoo, Queens Zoo, and New York Aquarium all host the World Conservation Society.[10] This is a synergistic relationship between a leading wildlife conservation organization and New York City's aquarium and zoos. The World Conservation Society focuses on at-risk birds, amphibians, and mammals in fifteen regions, including temperate Asian mountains and grasslands, Patagonia in South America, and Madagascar and the western Indian Ocean.[11]

The Monterey Aquarium teaches the teachers. Their teacher development programs support climate action summits, help instructors get students excited about protecting coastal ecosystems, develop awareness about ocean plastics, and connect conservation and technology.[12]

The core mission of zoos and aquariums remains to inform and inspire people. Each year, the Association of Zoos and Aquariums' Saving Animals From Extinction (SAFE) program spotlights ten endangered species. SAFE engages some 180 million annual visitors and obtains corporate sponsorship to help protect habitats, mitigate threats, and restore threatened populations. SAFE species include the African penguin, the Atlantic coral, the western pond turtle, and the black-footed ferret.[13]

In addition, some young people volunteer at zoos and aquariums. I know a young woman who personally introduced the world of butterflies to zoo visitors. How better to learn about ecology while creating Handprints?

NATURE SCHOOLS

There are schools dedicated to introducing people to nature.

Forest schools, also called "forest kindergarten" and "nature pre-schools," serve children three to six years old. Originating in Europe, nature-based preschools typically take children on weekly visits to the local nature preserve. The kindergarten in Quechee, Vermont, dedicates one day a week to learning in the woods—Forest Friday. In Bethel, Vermont, class begins on Monday morning with a one- to two-hour hike, weather permitting. Children's improved social and problem-solving skills helps them meet academic goals.[14] The American Forest Kindergarten Association shows dozens of member schools across the United States and Canada on their member map.[15]

On November 8, 2016, two-thirds of Oregon voters chose to offer fifth and sixth graders Outdoor School. Each year about fifty-five thousand students participate.[16] Portland schools first offered Outdoor School in

1957. For one week during the school year, students learn about nature firsthand. Besides raising academic performance and reducing high school dropout rates, Outdoor School opens the door for environmental inspiration.[17] An eleven-year-old girl who participated in the program said, "You can actually feel nature. It's not just saying, 'This is what a fern looks like.' You can actually feel the fern for yourself."[18]

An Apache Indian by the name of Stalking Wolf taught two young boys about tracking and survival in the Pine Barrens of New Jersey. One of those boys, Tom Brown Jr., authored *The Tracker* and formed the Tracker School in New Jersey in 1978.[19] More than a hundred different classes are based on the foundations of primitive skills and spiritual knowledge and teach students to "survive in any environment lavishly."

Brown, in turn, inspired other nature schools. Through Deerdance in Oregon, Terry Kem helps people develop "inner and outer awareness, and other skills needed for reconnecting and communicating with the natural worlds."[20] Also in Oregon, Trackers Earth, led by Tony Deis, offers a three-season immersion program in which students learn to make fire, natural shelter, shoes, and a kayak, among other things. One Trackers Earth assignment is to spend fifteen minutes in the same spot in nature every day for thirty days. Wilderness Awareness School, located near Duvall, Washington, offers a two-year Anake (protector) program that graduates people who go on to form other nature schools. Wilderness Awareness also hosts a self-guided Kamana Naturalist Training program you can do at your own pace and place.[21]

Nature is the setting for all eleven Outward Bound schools. Every year some thirty-five thousand Outward Bound students get in touch with nature at a visceral level.[22] Similarly, since 1965, the National Outdoor Leadership School has given us 280,000 graduates who respect the wilderness through their experience with backpacking, canoeing, kayaking, rock climbing, skiing, and wilderness medicine.[23]

ENVIRONMENTAL COLLEGE EDUCATION

Much has changed since I entered college in 1968. At that time, few if any schools offered environmental degrees, partly because the subject was too "cross-disciplinary." My alma mater, Arizona State University, now offers more than ninety-five graduate-level environmental life science doctorate course options, including Ecological Climatology and Geomicrobiology.[24]

Today, unique ecology curriculums are scattered across the map. As a project, students at Vermont's Middlebury College started 350.org with Bill McKibben, author of *The End of Nature*. At the College of the Atlantic, in Maine, every undergraduate earns a bachelor of arts in human ecology, the study of the relationship between humans and their environments. Off-campus internships range from a summer in the Arkansas River National Wildlife Refuge to a winter with Sustainable Bolivia.[25] Arizona's Prescott College teaches environmental education in collaboration with partners in Norway and Mexico.

In 1968, Erik Forsman, a graduate student at Oregon State University, studied the spotted owl's relationship with temperate rainforests. The spotted owl went on to become the indicator species that eventually brought ancient forest logging to a halt.[26] The University of Oregon, Portland State University, and Oregon State University partner to allow environmental studies graduate students to craft interdisciplinary projects and pursue concurrent degrees in law, psychology, geography, and policy studies.[27]

The majority of all life on earth, 50 to 80 percent, lives in the ocean.[28] A degree or two in marine science or oceanography may be appropriate for those who love the sea. The University of California San Diego hosts the Scripps Institute of Oceanography, which claims to be the "preeminent center for ocean, Earth, and atmospheric research, teaching, and public education."[29] Dalhousie University in Halifax, Nova Scotia, hosts the Ocean Tracking Network, which comprehensively examines ocean conditions and marine life.[30]

When I helped the Portland Water Bureau select a sustainability specialist, forty people were deemed "qualified." We reviewers determined twenty of those candidates to be "highly qualified," having both relevant degrees and experience. Unfortunately, only one person would get that particular job. The good news is that we now have a wealth of talent willing and able to heal our planet.

HANDPRINT OPPORTUNITIES

8.1 Tell environmental stories to your children. The stories we grow up with shape our outlook on the world. When in doubt, check with your local librarian.

8.2 Insist on environmental literacy for your child. Find out if and how your local school teaches ecology. Find out about your state's environmental education plans and standards from the North American Association for Environmental Education.

8.3 Help your local school become a Green and Healthy School. For information, check out the Green and Healthy Schools Academy online.[31]

8.4 Visit or volunteer for a botanical garden, aquarium, or zoo. Take your children and invite your friends' children. Volunteer with these institutions for hands-on experience for an eye-opening experience. The Phipps Conservatory and Botanical Garden in Pittsburgh boasts fourteen interconnected greenhouses with twenty-three distinct gardens and periodically changing exhibits. Their Center for Sustainable Landscapes is a zero net energy building that captures and treats its sanitary and stormwater on-site.

8.5 Support efforts to save endangered species through your local zoo or aquarium. Institutions accredited by the Association of Zoos and Aquariums support over 375 conservation projects worldwide.[32]

8.6 Send someone to a nature school. Attend one yourself.

8.7 Support sustainable education services. These include your local recycling hotline, the Greater Portland (Oregon) Sustainable Education Network, and the Sustainable Agriculture Research and Education program hosted by the University of Maryland.[33]

8.8 Translate good intentions into a career. Consult A Guide to Green Careers by Michael Hoffman online. It even includes scholarship opportunities.[34] Consider environmental science or environmental engineering.

8.9 Foster indigenous ways of learning within your child's school system. For guidance check out *Contemporary Studies in Environmental and Indigenous Pedagogies*, edited by Andrejs Kulnieks, Dan Roronhiakewen Longboat, and Kelly Young.[35]

IN YOUR JOURNAL

How will *you* help teach a young person about ecology?

9

NURTURE A LIFELONG
RELATIONSHIP WITH NATURE

In *Last Child in the Woods*, Richard Louv makes a compelling case that we need to get our young people in touch with nature. He reports declining attendance at national parks and laments the increasing quantity of video time in young people's lives. He worries about the paucity of young people inspired to become activists and stewards of the future.[1] Maria Alice Campos Friere says attention deficit disorder is really a "nature deficit disorder."[2]

EXPOSING YOUNG PEOPLE TO NATURE

A child's nature education need not depend on an enlightened school system or a hospitable political climate.

The American Camp Association (ACA) lists 2,400 camps as members. The ACA says, "We seek and appreciate what is real, genuine, and non-artificial. In seeking those qualities in people as well as in the actual world, we foster understanding of the importance of human connections for survival and of the critical connection to our physical world. Campers realize the need to protect not only one another, but also the environment in which they live."[3] The Missouri Ozark Mountains hosts YMCA Camp Lakewood and its wilderness ranger program. Camp Lakewood provides accommodations for young people with AIDS, asthma, cancer, and cystic fibrosis.[4] Every summer, The Fresh Air Fund hosts three thousand youngsters at one of six camps north of New York City.[5]

A young person can come to love nature in many ways. I went fishing with my dad in northwest Pennsylvania's wetlands and hunting with my grandfather in Arizona's high country. My first five-mile hike in the desert south of Phoenix earned me the Tenderfoot rank in Boy Scouts.[6] My high school Explorer post took me to Havasupai Canyon. In college, I found camaraderie with the Arizona Mountaineering Club.

For some, city parks may be the only available exposure to nature. The Portland (Oregon) Parks Environmental Education program offers field trips and restoration opportunities for young people.[7]

In the United States, forty-five million people (age six and over) went on at least one hike in 2017, 14 percent of our population.[8] How can the rest of us get in touch with nature?

ECOTOURISM

I like to help people move from one house to another. In return I get exercise, camaraderie, and appreciation. The same can be said, on a much larger scale, for ecotourism.

Ecotourism in its various manifestations offers great opportunities to learn by immersion in the active preservation of the environment. *Ecotourists Save the World* by Pamela Brodowsky lists over three hundred volunteer experiences to conserve, preserve, and rehabilitate wildlife and habitats. The ones listed in the United States (nearly half the book) are generally "free in exchange for services." Some even provide housing arrangements. How would you like to spend a few weeks in Idaho's Deer Flat National Wildlife Refuge or Mississippi's Sandhill Crane National Wildlife Refuge?

Megan Epler Wood's initiative led to the founding of the International Ecotourism Society (TIES) in 1990.[9] The TIES motto is "Uniting conservation, communities and sustainable travel." TIES provides training, certification, and advocacy for innkeepers, guides, and government officials in 135 countries. TIES initiatives are unfolding in such countries as Kazakhstan, Montenegro, and Vietnam.[10]

Costa Rica thrives on ecotourism with twenty national parks that protect 5 percent of the earth's biological diversity. You can witness green sea turtle nesting sites at Tortuguero National Park, touch 140 species of trees at Cabo Blanco National Reserve, and (try to) see 240 bird species and 40 mammalian species at the Tapanti-Macizo de la Muerte National Park.[11]

Catering to ecotourists offers developing communities a more sustainable business model than resource depletion. International ecotourists pay upward of $1,000 to $4,000, plus transportation, to visit once-in-a-lifetime destinations. The money keeps underfunded conservation efforts viable. To wit: the Tanzania African Wildlife Conservation Program, the Andean

Conservation Project, Tracking Baja's Black Sea Turtles, the Crocodile Conservation Program in India, and the Orangutan Project in Indonesia.[12]

EXPERIENTIAL EDUCATION

Some of our environmental heroes claim their genius outside the box of formal institutions.

Naturalist John James Audubon (1785–1851) left a powerful legacy thanks in part to his youthful interest in birds and nature. He learned from a long series of mentors. From Charles-Marie D'Orbigny, in France, young Audubon learned taxidermy and scientific methods. He was inspired by poet-naturalist Alexander Wilson. Audubon learned to paint from landscape artist John Stein and history artist Thomas Cole. This life of attention to science, art, and nature still inspires today's Audubon Society.[13]

John Muir dropped out of the University of Wisconsin; yet he became the nation's foremost advocate for national parks. His education in botany was fostered by a walk from Indiana to Florida, making sketches along the way. His assertion that the walls of California's Yosemite Valley were formed by glaciers came from close observation while working and hiking in the nearby mountains.[14] Muir's eloquent writing about the area now known as Glacier Bay National Park can be traced to his participation in the 1899 Harriman Alaska Expedition.[15]

Aldo Leopold, in his book *A Sand County Almanac*, tells of a woman in Ohio who carefully observed the habits of song sparrows in her backyard. In so doing, she discerned the fundamentals of "sparrow politics, sparrow economics and sparrow psychology."[16] She was a citizen scientist.

Every year some 200,000 people participate in citizen science projects sponsored by Cornell University. These include Project Feederwatch, NestWatch, and the Great Backyard Bird Count. A joint project of Cornell Lab of Ornithology and the Audubon Society called eBird amasses millions of real-time bird observations.[17]

Oxford University refused to allow Amory Lovins to pursue a doctorate in energy, so he dropped out and worked for Friends of the Earth, forming its United Kingdom chapter. In 1977, Amory Lovins wrote as profound a work as any dissertation, *Soft Energy Paths: Towards a Durable Peace*. His book established, in the arenas of academia, utilities, and government, the principle that energy conservation can reduce or eliminate the need for new centralized power plants.[18]

Oregon's Lloyd Marbet led a successful eighteen-year (1976 to 1993) effort to end nuclear power generation in Oregon. He dropped out of community college before leaving for the Vietnam War. After returning, Marbet read *Perils of the Peaceful Atom: The Myth of Safe Nuclear Power Plants* published in 1970 by Richard Curtis. A motivated Marbet attended sitting hearings, filed court motions, and collected thousands of signatures for statewide ballot initiatives. His commitment inspired allies—including me.

After leaving full-time employment in 2003, I began to yearn for a PhD. I even found a school that would let me pursue a cross-disciplinary degree in sustainability and spirituality. Yet I could not justify the expense when I had no plans to pursue high-paying employment. Besides, I noticed that most of my heroes were self-directed learners.

Hearing about this dilemma, my wife suggested that I instead pursue a "PhD from the universe rather than the university." Suddenly, my adventures in sustainability began to feel like experiential coursework: Our eco-retrofit home. Consulting for nonprofit environmental organizations. Choosing to live without a car. Voluntary carbon offsetting. The list goes on. My blogs are like class reports. The talks I give are akin to seminar presentations. As for research, I am developing the Handprint.

Several of my sources have also chosen to go deep in their field rather than academia. Addie Fisher translated her study of architecture and interior design into her blog *Old World New.*[19] Her posts include "How to Create a Paperless Kitchen" and "Ultimate Guide to Sustainable Clothing Brands for Adults." English farmer Ben Eagle writes an ongoing series of Meet the Farmer posts on *Thinking Country.*[20] Sandor Katz, in publishing *The Art of Fermentation*, has become a go-to resource for fermented foods (chapter 11). Rob Greenfield raises sustainability awareness with projects such as living in a tiny house, growing and foraging his own food, and wearing all his trash. Paul Stamets studies mycelia (mushrooms) with the passion of a prophet. In doing so, he allies us with the natural world in cleaning up toxic messes (chapter 15).

A note of reassurance for academics. A formal degree has value too. It assures universities, employers, and clients that the holder has met the rigor of an independently accredited program.

HANDPRINT OPPORTUNITIES

9.1 Go camping and hiking with your children. Take the neighbor children too.

9.2 Help your child participate in nature-related activities. Camp Fire invites *all* young people to participate in camps, service projects, and after-school activities. Over four hundred National Park Service sites offer Junior Ranger programs. The National Wildlife Federation has Ranger Rick adventures for wherever you live. Also, check out your REI Co-op events calendar for classes and day trips.

9.3 Closely observe a tree over the seasons, or the neighborhood dynamics over a month, or the waste stream of a local school. Tell a child what you see. Tell the school board. Tell the neighborhood association.

9.4 Become an ecotourist. Support good stewardship while you are at it. Check out *Ecotourists Save the World* for ideas.

9.5 Embrace citizen science. Join an organization such as Cascadia Wild, which tracks wildlife on Oregon's Mount Hood. Check out *Citizen Scientist: Searching for Heroes and Hope in an Age of Extinction* by Mary Ellen Hannibal.[21]

9.6 Study with a mentor. Ensia offers mentorships for environmental storytellers in mediums such as articles, videos, graphics, and photos.[22] The Society of Environmental Journalists has a mentorship program (just sign up).[23] The Institute for Tribal Environmental Professionals at Northern Arizona University offers mentoring in managing tribal waste.[24] The nonprofit Think Beyond Plastic provides mentors and innovation hubs to accelerate "innovation for a world free of plastic pollution."[25]

IN YOUR JOURNAL

How will *you* deepen your relationship with nature?

IV

FOOD HANDPRINTS

10

HONOR THE SOIL

There is nothing more fundamental to our sustainability than soil and how we grow food. Agriculture, when you include the industrial manufacture of fertilizer and moving food to market, is second only to power generation in contributing to global warming—ahead of transportation.[1] Agriculture as we've come to know it uses 70 percent of the world's fresh water. The discerning eye can see more than one story of agriculture unfolding simultaneously.

A Dickens novel view of agriculture sees a profound consolidation of farm acreage following in the wake of retiring farmers whose land is no longer worked by their children.[2] Walking through the Midwest on the Compassionate Earth Walk in 2013, I saw innumerable abandoned schoolhouses. The child in school who says he or she wants to grow up to be a farmer may be laughed at by his or her peers.

However, since 2000, the number of farmers planting fifty acres or less has increased.[3] These are often part-time farmers, specialty farmers, farmers willing and able to use more labor-intensive and less chemical-intensive practices. *Some farmers, ranchers, and scientists, through a variety of regenerative agricultural practices, have learned how to reverse the flow of carbon back into the soil.* This is the fertile ground for a sustainable agricultural revolution.

ORGANIC AGRICULTURE

Our journey into agricultural hope starts with organic farming. As of 2018, 6.5 million acres (2.6 million hectares) met the organic certification requirements of the US Department of Agriculture, up from 5.3 million acres in 2015.[4] This cropland and rangeland is managed by more than 17,600 "operators," up from 15,000 in 2015.[5]

The roots of organic farming can be found in the biodynamic movement, a collaboration between Dr. Rudolf Steiner and farmers in Germany during the early twentieth century.[6] Treating the farm as an ecosystem, including using animal manure to fertilize the soil, responded to the industrial farming of the day. The practice of intentionally letting a field lie fallow, much as practiced in biodynamic farming, is literally gaining ground with large-scale agriculture. A periodic no-till season gives microbial soil inhabitants relief from the heavy wheels of farm machinery. Meanwhile, natural plants (weeds) can revitalize the soil.

In 1940, Walter James, an early student of biodynamic farming in England, wrote *Look to the Land*. In it, he coined the term "organic farming" from the concept of the farm as an organism.[7]

Certified organic land now falls under the rubric of the Organic Foods Production Act of 1990. Organic certification is the best and the worst of government bureaucracy, driven by activism and science. The Code of Federal Regulations pertaining to organic food and fibers reads thick and dense—eighty-one provisions.[8] Farmers complain for good reason. And we consumers really can be assured that what we buy is eco-friendly.

CODE OF FEDERAL REGULATIONS PERTAINING TO ORGANIC FOOD AND FIBER (SELECTED PROVISIONS)

Land requirements. Any field or farm parcel . . . represented as "organic" must . . . have had no prohibited substances . . . applied to it for a period of 3 years immediately preceding harvest of the crop.

Soil fertility and crop nutrient management practice standard. . . . The producer must manage plant and animal materials to improve soil organic matter content in a manner that does not contribute to contamination of crops, soil, or water by plant nutrients, pathogenic organisms, heavy metals, or residues of prohibited substances.

Livestock Living Conditions. The producer of an organic livestock operation must establish and maintain year-round living conditions which accommodate the health and natural behavior of animals.

Facility pest management practice standard. If enumerated natural pest management practices have been exhausted . . . The organic plan must include a list of all measures taken to prevent contact of the organically produced products or ingredients with the substance used.

On-site inspections . . . A certifying agent must conduct an initial on-site inspection of each production unit, facility, and site that produces or handles organic products and that is included in an operation for which certification is requested . . . annually thereafter . . . may be announced or unannounced.

Record keeping by certified operations. A certified operation must maintain records . . . for not less than 5 years.

Source: National Organic Program, "Electronic Code of Federal Regulations," https://www.ecfr.gov/cgi-bin/text-idx?tpl=/ecfrbrowse/Title07/7cfr205_main_02.tpl.

Organic farms come in all sizes and specialties. Tuscarora Organic Growers is a cooperative of more than forty organic farms near Hustontown, Pennsylvania. They grow over 1,200 different produce items, selling to Whole Foods markets as well as their local community-supported agriculture and farmers markets.[9] Earthbound Farm, started by Drew and Myra Goodman, grew from 2.5 acres in 1984 to more than 30,000 certified crop acres in 2020.[10] Horizon Farms, now owned by White Wave Foods, partners with more than six hundred family farms. The Horizon Organic Producer Education program has helped many farms make the transition to organic and has provided over $120,000 in college scholarships to future farmers.[11]

Between 2013 and 2018, US organic food sales grew nearly 10 percent per year.[12] As of 2018, organic farmers and ranchers serve a $52.5 billion US organic market ($47.9 billion food; $4.6 billion nonfood items such as lotions, cloth, and alcohol). This accounts for 5.7 percent of all food sold in the United States.[13]

Increased health concerns during the COVID-19 pandemic drove a 20 percent increase in organic fruit sales in the spring of 2020.[14] This shift portends a long-term overall increase in organic sales.

AGRICULTURAL CARBON SEQUESTRATION

Recent science validates the thesis that healthy soil is a living system, not just a collection of chemicals. *The Soil Will Save Us*, by Kristin Ohlson, describes a network of symbiotic relationships. Plants provide carbon "exudes" to bacteria and fungi. Those microbes pull mineral nutrients from sand and rock. Nematodes and worms eat the microbes and release (poop) the nutrients in forms the plants can use.[15] Up to 72 percent of the carbon captured by a plant can be held in the soil.[16] Along the way, aggregates build up in the soil, which keeps it soft and able to hold water.[17]

This complex process is the essence of soil carbon sequestration. It all happens naturally, if all the players are present and free to do their work.

Conversely, today's prevailing agricultural paradigm fails to sequester carbon. Agricultural production produces 11 to 15 percent of anthropocentric (human-related) carbon emissions.[18] (This does not include carbon loss from deforestation and land clearing primarily for agricultural purposes—another 15 to 18 percent.) Tilling physically breaks up the soil's complex and delicate structure. Heavy equipment compacts it. Pesticides and chemical fertilizers create a hostile environment for the micro fauna and flora. Monocultures reduce the diversity of micropredators that consume pests.

Judith Schwartz, in *Cows Save the Planet*, loosely summarizes various soil sequestration strategies in the phrase "oxidize less; photosynthesize more."[19] In other words, send carbon deep into the soil and avoid disturbing it while we raise our food. A growing cadre of dedicated researchers and farmers around the world has been doggedly discovering, creating, and demonstrating carbon sequestering agricultural systems that meet the needs of today's farmer, today's market—and the microfauna in the soil.

Regenerative agriculture, as this movement of experience and expertise is often called, intends to restore carbon that was lost from the soil through generations of clearing and plowing.[20] Abe Collins in Vermont revitalizes farms, drawing on some forty regenerative agricultural strategies.[21] Farmland LLP, based in Oregon, buys worn-out farms to transition them into certified organic farms.[22] Ademir Calegari of Brazil advises farmers to plant a "cocktail" of cover crops that feed the microorganisms year-round.[23]

Dr. Christine Jones, who lives in Australia, researches the soil biology of carbon sequestration. She tells her audiences to move beyond reductionist science. The microbial communities in the soil are more than the sum of their molecules. The more diversity of cover crops on top of the soil, the more carbon sequestration below the soil.[24] Dr. Jones posts articles relating to regenerative agriculture on her Amazing Carbon website.[25]

Agricultural giants and associated government agencies have rarely funded research that supports regenerative agriculture. Most approaches to carbon sequestration reduce the need for industry-produced fertilizer, insecticides, and seed. Handprints often originate outside the "system."

In May 2018, the Regenerative Organic Alliance released its Framework for Regenerative Organic Certification. Its goals are "to improve soil health and organic matter over time and sequester more carbon, to improve animal welfare, and to provide economic stability and fairness to farmers, ranchers and workers." Leadership is provided by the Rodale Institute. Members of the alliance include organic stalwarts such as Horizon (dairies), Demeter (biodynamics), Patagonia (clothing), and Nature's Path Organic (grain-based foods). Meeting existing government organic labeling standards is a prerequisite for regenerative organic certification. This collaborative effort is a huge step forward in maturing and marketing the sustainable farming industry.[26]

Practices described below are options woven into to the regenerative organic criteria.

PERENNIALS

Crops that bear fruit and seed year after year without the need for replanting are called perennials. Perennial farming maintains the complexity and integrity of the soil, minimizes erosion, and avoids using energy to plow fields. Perennials sequester carbon.

We humans already have a collection of well-established perennials. Tree crops include olives, apples, citrus fruit, bananas, nuts, maple (syrup), coffee, cork, and rubber. Alfalfa, asparagus, grapes, and hay are perennials as well. Likewise, bamboo.[27] Eric Toensmeier, in *Perennial Vegetables*, documents over one hundred perennial food crops that are eaten somewhere in the world.

It takes time—sometimes decades—and resources to breed new strains of perennials that can economically displace carbon-profligate, university-researched annual crops. Also, the year or more before perennials can be harvested slows their introduction. That said . . .

In China, the Yunnan Academy of Agricultural Sciences is developing perennial rice.[28] (This complements a labor-intensive approach to growing *annual* rice in ways that reduce needs for water, chemical fertilizers, and pesticides, while releasing less methane. This is called the system of rice intensification, used by twenty million farmers in Asia.)[29]

The Land Institute, founded by Wes Jackson in Kansas, is developing perennial strains of wheat, wheatgrass, sorghum, and sunflowers. Sorghum is used as cattle feed and ethanol for fuels. The Land Institute is also developing "natural system agriculture" in a "polyculture" setting. They plant new strains of wheat together with other plants, like legumes and oilseeds. Thus, they challenge the monoculture paradigm and avoid replowing the ground.[30]

INTENSIVE RANCHING AND FARMING

Alan Savory, a field biologist, painfully observed that the land in Africa seemed healthier when elephants were tromping on it than after they had been killed—to protect the land. This led to a decades-long odyssey to discover what a heathy landscape really needs.

In temperate grasslands with long periods of low humidity, *four* times as many domestic animals can be supported on a piece of land using holistic management. Cattle, goats, sheep, camels, and horses are moved regularly as a herd rather than leaving them to graze randomly. Humans essentially play the role of displaced predators, such as lions. Divots that hooves make catch what little rain falls. Dung and crushed grass ("litter") prevent the water from evaporating while fertilizing the grass. How often the herd passes through depends on how fast a given kind of grass grows.[31]

Savory posits, eloquently, that this way of ranching is key to stopping desertification around the world. He has demonstrated his methods in Zimbabwe, Mexico, and Argentina among other places. He even asserts that regularly plowing grass back into the soil is a practical way to remove carbon dioxide from the atmosphere—carbon sequestration. Judith Schwartz, referring to Savory's principles, went so far as to title her book *Cows Save the Planet*.[32]

Joel Salatin and family's Polyface Farm, in Virginia, also moves animals to achieve extraordinary production. The key is short periods—a few days—of intensive grazing followed by moving the herd to a new location. Both his beef and his pastures thrive. This is how the buffalo moved in the Great Plains. Salatin uses the same principles to raise chickens using rolling coops. Salatin's books include *Pastured Poultry Profits* and *Everything I Want to Do Is Illegal*.[33]

Caution. Intensive agriculture requires labor to keep the herd moving and maintain fencing.

AGROFORESTRY

Agroforestry adds trees to the mix of plants, yielding value in a managed farm or ranch.[34] Trees sequester carbon both above and below ground.

When a tree is used for building material, carbon is then sequestered in that material for the life of the structure. All agroforestry approaches sequester carbon.

AGROFORESTRY PRACTICES PROMOTED BY THE UNIVERSITY OF MISSOURI

- **Windbreaks** control soil erosion, provide wildlife habitat, and protect livestock. The United States planted millions of trees for windbreaks in the 1930s. In the United Kingdom, trees and shrubs between fields and along roads are called hedgerows.
- **Riparian buffers** are intentionally wild zones in an otherwise managed landscape. They offer ecological services such as flood control, shade, water filtration, and wildlife habitat.
- **Forest farming** promotes the growing of high-value shade crops, like shiitake mushrooms, while growing trees to harvest for lumber.
- **Alley cropping** intersperses rows of annual crops with rows of trees, maximizing tree growth and diversifying farmer income.
- **Silvopasture** integrates livestock management with trees that can produce a crop or fodder. For example, in Spain, cattle graze among oak trees.[a]

Source: "What Is Agroforestry?," Center for Agroforestry, University of Missouri, accessed July 13, 2017, http://www.centerforagrofor estry.org/practices.

---

[a] Eric Toensmeier, *The Carbon Farming Solution* (White River Junction, VT: Chelsea Green, 2016), 41.

Intensive silvopasture integrates intensive ranching (described above) with tree fodder. As practiced on half a million acres (two hundred thousand hectares) in Australia, cows rotate between fenced areas while they browse on the leaves of the fast-growing leucaena tree (known in different places as white leadtree, jumbay, river tamarind, and white popinac). Leucaena, with its tendency to crowd out other plants, is considered an invasive species in some areas.[35] That said, the intensive silvopasture strategy offers perhaps the *highest potential for agricultural carbon sequestration*.[36] Compared to conventional grazing practices, silvopasture yields two to ten times the meat production and lowers methane emissions.

You may not find tropical homegardens in an agricultural university handbook. Homegardens (distinct from backyard home gardens) are multilevel combinations of various trees and crops, often grown in association with animals around a homestead. In Indonesia, tropical homegardens grace 31.1 million acres (12.6 million hectares). These complex ecosystems can also be found in South Africa, Guatemala, Bangladesh, and Sri Lanka. Since homegardens manifest generations of learning about what works in a given area, they are hard to replicate.[37]

HANDPRINT OPPORTUNITIES

For Everyone

10.1 Support The Land Institute in developing natural system agriculture, especially perennial crops.

10.2 Buy agricultural carbon offsets that compensate farmers and ranchers. Climate Trust lists three approved types of agricultural carbon sequestration: changes in fertilizer management, rice growing management systems, and soil carbon building.[38] The maturing of regenerative agriculture should yield more carbon offset opportunities.

10.3 Spread compost on bare or compacted land to revitalize its microbial community and sequester carbon. Use a mix of green matter (such as vegetable kitchen scraps) and brown matter (such as fallen leaves and paper). Do it in your backyard. Do it on a ranch, as Whendee Silver and her UC Berkeley students did in California.[39]

10.4 Study botany and indigenous food cultivation. Let both science and the wisdom of our elders inform our farming.

For Farmers

10.5 Use sustainable farming practices. Start with the Sustainable Agriculture Research and Education program to learn about crop rotation, no-till farming, hoop houses, developing a transition plan to organic farming, and much more.[40]

10.6 Become certified as an organic grower. Assure distant consumers that the land is respected.

10.7 Practice "mob grazing" so that animals take care of the soil—and sequester carbon.

IN YOUR JOURNAL

How will *you* honor the soil?

11

PARTNER WITH
EARTH-FRIENDLY FARMERS

When my children were growing up, we held hands for a "moment of silence" before meals. I saw it as a way to allow anyone to pray—or not—in any way he or she wanted. Now I see that, in a modest way, we were making the very act of eating sacred. Food is fundamental to life. In fact, our culinary practices can, if we are intentional, contribute to the health of the planet. We can, as Anna Lappé says in *Diet for a Hot Planet*, claim "the power of the fork."

GARDENING

Every time my wife and I move, we convert our lawn into a healthy, productive landscape. The process usually starts with sheet mulching the grass with cardboard and wood chips—no need to dig up the turf. Then we plant gardens, berry bushes, and fruit trees. The result, in five years or so, is a food forest! My wife, our grandsons, and I have had success with cucumbers, plums, quince, raspberries, strawberries, green beans, tomatoes, lettuce, kale, mulberries, persimmons, and, of course, zucchini.

Driving down the street, we see neighbors also repurposing their lawns. Around a third of US households grow some of their own food.[1]

For those without a yard, the community garden is often a viable resource. The Trust for Public Land reports that there are more than twenty-nine thousand community gardens in US city parks.[2] The American Community Gardening Association maintains an online repository of "green docs." Titles include "Tips for Urban Gardeners," "A Gardening Angel's How-To Manual: Easy Steps to Building a School Garden," "New Farmers Training Manual," and "Why Is Compost Special."[3] Camaraderie with fellow gardeners can be one of the greatest benefits of a community garden.

FARMERS MARKETS

Farmers markets are making a comeback. Farmers markets greatly reduce the 1,500-mile transportation Footprint associated with store-bought food. They also virtually eliminate packaging.

Outside our local People's Food Co-op, at the Wednesday farmers market, we simply ask the seller how he or she grows food without the intervention of a bureaucracy. Odds are, even if you do not ask, the food will be more local and more organic and less genetically modified than at the supermarket.

Farmers markets have been around for thousands of years. However, the commoditization of food, starting in the late 1800s, and the advent of supermarkets in the 1950s marginalized farmers markets—until recently. In 1994, there were 1,755 farmers markets listed in the National Directory of Farmers Markets. As of 2019, 8,790 markets are listed. That is a fivefold increase, or 20 percent a year.[4]

COMMUNITY-SUPPORTED AGRICULTURE

Farmers and customers can cooperate over time through community-supported agriculture (CSA). The CSA movement was pioneered in Europe in the 1960s by biodynamic farmers and women's groups who were community partners. They were inspired by scientist and philosopher Rudolf Steiner, who taught that a network of human relationships should replace systems of employers and employees.[5] In the mid-1980s, Jan Vander Tuin brought his experiences in Switzerland to the United States and helped found the Indian Line Farm in Massachusetts. About the same time, Trauger Groh came from Germany to be with his new wife and helped found the Temple-Wilton Community Farm in New Hampshire. One of the tenets we draw from these farms is that a land trust ensures that the acreage will continue to be available for farming without the farmers having to be wealthy enough to own it. Robyn Van En, a cofounder of the Indian Line Farm, founded CSA North America to help develop CSAs across the continent.[6]

Several different CSA arrangements have evolved. One model has subscribers who share both the bounty and some of the risk associated with the harvest. Sometimes a parent organization with a social agenda, such as the Hartford Food System (operating Holcomb Farm) in Connecticut, makes the fundamental decisions. Occasionally, growers with a common heritage band together to operate a farmers market. Raíces Cooperative Farm serves the Hispanic community near Hood River, Oregon.

In 2006, the US Department of Agriculture (USDA) listed 761 CSAs in its database.[7] Up until 2019, the Robyn Van En Center at Pennsylvania's Wilson College maintained a CSA database with over 1,650 growers from every state.[8] LocalHarvest tracks upward of thirty thousand farms, farmers markets, restaurants, and grocery stores that feature local foods in its directory. LocalHarvest also offers software to help manage a CSA.[9]

If you live near a coast, consider a community-supported fishery (CSF). Massachusetts's Cape Ann Fresh Catch CSF and Maine's Port Clyde Fresh Catch CSF offer individuals and restaurants a set number of pounds of local fish and shellfish every week. The buyers pay about the same as at the store; the local fishermen make six times as much.[10]

FOOD LABELING

Information is power. When we know what is in our foods, we can better choose products that support our values. Access to information on quantity, calories, nutrition, point of origin, farming methods, and the use of genetic engineering is, or can be, a civil right.

The basic product labeling requirements we in the United States are familiar with were established under provisions of the Fair Packaging and Labeling Act signed by Lyndon Johnson in 1966. Since then, labels must identify the product; provide the name and place of business of the manufacturer, packer, or distributer; and state the net quantity of the contents.[11] The same law that regulates organic farming (chapter 10) regulates organic labeling. The legal terms "100 percent organic" and "organic" (95 percent) and "made with organic ingredients" (70 percent) are used in labeling. By taking these words and the accompanying label seriously, we empower ourselves to partner with earth-friendly farmers. My wife will go without rather than buy vegetables without an organic label.

Similar labeling provisions are now in place in the European Union, Canada, Australia, Japan, and India.[12]

Regional sustainable food labels can also offer the buyer assurances. Some farms in New Hampshire use the Certified Naturally Grown label, mirroring federal organic farming practices with little paperwork and annual peer farmer inspections. This allows for smaller-scale operations and potentially interim certification before the multiyear process to achieve USDA certification is completed.[13] The Montana Sustainable Growers Union offers the Homegrown label with similar assurances.[14] Oregon Tilth offers "transitional certification" for farms that are partway through the process of qualifying for full organic farm status.[15]

Due to the many comments received during the USDA rule-making process, the official definition of organic food does *not* tolerate genetically engineered (GE) ingredients. In May 2015, the USDA signaled its intent to endorse a voluntary Non-GMO/GE Process label on commodity products. Meanwhile, the Non-GMO Project offers a butterfly label, which verifies that "a product has met rigorous best practices," including "ongoing testing, and has passed third party audits and inspections for GMO avoidance." The Non-GMO Project has verified over fifty thousand products.[16]

Vermont was the first state to require GE labels—as a result of Act 120, signed into law on May 8, 2014.[17] So far England, Canada, Mexico, and the United States do not require such labeling at the national level. However, as of 2014, sixty-one countries do require that food labels inform buyers of any GE content. These include Finland, Japan, Russia, Ukraine, Kenya, and Ethiopia.[18]

EATING AS IF THE ENVIRONMENT MATTERED

Here are some strategies to create Handprints with the food we eat.

Frances Moore Lappé (mother of Anna Lappé, who was mentioned above) reintroduced us to the idea that we can eat healthy without meat in her 1971 book *Diet for a Small Planet*. (Many Hindus, Jains, and Buddhists have eaten vegetable-based diets for centuries.) Vegetables require less water during production than meat and result in lower carbon emissions. Besides, vegetables are healthier, and eating a vegetarian diet avoids the degrading treatment of animals. Meat requires six to seventeen times more land than soybeans to produce an equivalent amount of protein, depending on the type of meat. Likewise, meat requires six to twenty times as much fossil fuel and four to twenty-six times as much water.[19] It is kinder to our fellow planetary inhabitants when we "eat lower on the food chain."

Raw foods avoid the environmental impact associated with processing plants and cooking, while retaining more vitamins and reducing obesity. Think of the environment as that extra incentive we need to try green smoothies. Robyn Openshaw, author of *Green Smoothies Diet*, tells us how.[20]

Naturally fermented food, such as sauerkraut, kimchi, and vinegar were traditionally kept without energy-consuming refrigeration. *The Art of Fermentation* by Sandor Katz tells us how to create fermented foods and why this is a revolutionary act.[21] Yes, there are health benefits. Fermented foods revitalize our digestive and immune systems (as long as commercial pasteurization is avoided).[22] Willow and I make sauerkraut in our glazed ceramic crock, three heads of cabbage at a time.

Canning bridges the gap between the abundance of harvesttime and the rest of the year—without the need for refrigeration. Today people in factories do most of our canning. From an environmental point of view, canning is much better than long-term refrigeration.

Dry food also avoids refrigeration. Squirrels and ants long ago taught our ancestors how to keep seeds dry. For millennia, people have been keeping food "squirreled away" in dry places. Root cellars keep potatoes and winter squash fresh for months. Gallon jars keep our rice and beans dry and available.

Bulk foods require less packaging, benefiting landfills and oceans. The Bulk Is Green Council seeks to (re)mainstream bulk foods.[23] Fresh Market #105 in Miami Beach, Florida, won the national 2013 Bulk Retailer of the Year Award. Bulk foods in small quantities may not yield consumer cost savings because dispensing the food still requires handling. Customers need to provide their own reusable containers to reap the no-packaging benefit.

Just as organic purchases protect the soil from chemicals and pesticides, purchase of regenerative foods fosters a healthier agricultural system: for the soil, the animals, the workers, and the climate. As supermarkets meet the demands of their eco-conscious customers, farmers gain the means to reverse the flow of carbon into the atmosphere. In the meantime, buy directly from farmers that practice silvopasture, intensive farming and ranching, and biodynamics (see chapter 10).

In particular, let's create a demand for perennial food—food that can be harvested without breaking up or compacting the soil. Perennials may be a stretch for some short-order cooks, but they can be the basis for a new wave of eco-sensitive cuisines. Perennial fruits and tree nuts already grace our diet. Use them more. I like cashews in a stir-fry better than meat *or* tofu.

FOOD FOR THE EARTH-FRIENDLY CONNOISSEUR

Use perennials to meet a variety of dietary needs while supporting regenerative agriculture.

> **Basic starches:** Use bananas, plantains, and breadfruit rather than sweet potatoes and yams.
>
> **Balanced carbohydrates:** Use chestnuts, carob, perennial fruits, acorns, and nuts rather than wheat, rice, and potatoes.
>
> **Proteins:** Use perennial beans, nuts, and leaf protein concentrates rather than annual beans, chickpeas (garbanzo beans), and lentils.
>
> **Protein oils:** Use seeds, beans, and nuts rather than soybeans, peanuts, and sunflower seeds.
>
> **Primarily oil:** Use olive, coconut, avocado, oil palm, macadamia, and shea (butter from nuts of Africa's shea tree) rather than canola, poppy seed, maize, cottonseed, sunflower, and peanut.

Source: Eric Toensmeier, *The Carbon Farming Solution* (White River Junction, VT: Chelsea Green, 2016), 131.

Many cookbooks offer vegetarian and vegan (no animal products) cuisine. The days when one compromised taste to eat vegetarian or vegan are gone. We also need a library filled with cookbooks that emphasize perennials, one of the foundations of regenerative agriculture.

NOT ALL MEAT IS CREATED EQUAL

About 5 percent of people in the United States identify themselves as vegetarian or vegan.[24] The rest of us need to recognize that, from an environmental perspective, it makes a difference what meat we eat and how much.

Lamb and beef have the largest Footprints of any supermarket meat. The ruminant digestive systems of sheep and cows release large amounts of methane, which is much more potent as a greenhouse gas than carbon dioxide.[25] (In the first two decades after its release, methane averages eighty-four times the global warming impact as carbon dioxide, but it dissipates faster over the long run.)[26]

In the words of one marketer, "Goat is the other (sustainable) red meat."[27] People around the world raise and consume more goats than any other meat animal. Goats require less space than cows, and they naturally control weeds. Goat meat is much lower in fat than beef, 3 percent versus 18.8 percent. Goat milk is nutritious and easier for most to digest. Over a period of thirty years, US goat production increased eightfold—without the use of concentrated animal feeding operations.[28] However, goats are also ruminants with relatively high greenhouse gas production.[29]

Swine (pigs) produce twice as much meat as beef for a given amount of feed; the single-chamber pig stomach digests more efficiently than the four-chamber cow stomach. Pigs reproduce faster, and pigs release less methane.[30]

Chicken or fish production results in one-fourth the carbon-dioxide equivalent greenhouse gas emission compared to cows and lambs.[31] The crowded windowless buildings in which most meat chickens are farmed raises moral questions.[32] Fish farming is big business as well, and management issues abound related to fish diseases and pollution of the surrounding waters.[33]

Summarizing, given the realities of today's agribusiness, *cattle are especially hard on the environment*. Chicken and fish are better. Plants grown to directly feed people are better yet. Regenerative agriculture, if and when it is mainstreamed, could change this assessment.

PARTNERING WITH GROCERY STORES

Most people still buy food from grocery stores. How can stores help us create Handprints? How can they help farmers nurture the soil? While stocking organic food is necessary, that may not be sufficient to motivate a shift to more earth-friendly eating. A sense of partnership is needed. But what does that look like? Here are a few examples from my area.

New Seasons Market is a benefit corporation. That means that the organization as a whole is committed to having a positive impact on the community and the environment. To start with, New Seasons gives 10 percent of its profits to local nonprofits, including, a while back, the Earth and Spirit Council. Their staff volunteer for trail building and invasive plant removal projects. Food is donated to Meals on Wheels for shut-ins and seniors. Deliveries to each store from a distribution hub is provided by

B-Line, which uses electric cargo tricycles.[34] Individual stores offer tastings of apples (a perennial crop) from local organic growers.

Whole Foods offers cooking classes and locally made foods. I'm waiting for them to offer classes on perennial cooking.

People's Food Co-op is our go-to store for bulk products ranging from olive oil to Biokleen dishwashing powder. This co-op refuses to carry meat and fish. It hosts a farmers market on Wednesdays, which draws in new customers.

Safeway sells bus tickets and offers plastic film recycling—in addition to their O Organics brand foods. The fact that they sell other food to other customers does not invalidate how Safeway helps me be more earth-friendly.

Walmart carries single-serving six-packs of shelf-stable Horizon organic low-fat milk. This might seem like an example of excess packaging, but it helps me live without a refrigerator while still indulging in an occasional bowl of cereal.

"Trader Joe's products are sourced from Non-GMO ingredients."[35] It is part of their branding. Also, Trader Joe's discontinued use of potential reproductive system disrupters bisphenol A (BPA) and bisphenol S (BPS) from their cash-register receipts.[36]

Grocery Outlet gives discount prices on overstock organic cereals, canned goods, and brands I have never heard of.

Handprint thinking endorses products and services wherever they help us to be more earth-friendly. When we live our eco-values, stores and farmers will partner with us.

HANDPRINT OPPORTUNITIES

11.1 Grow food in a garden. Help children get their hands dirty—and love it. Give away the surplus, and get to know your neighbors.

11.2 Buy food at farmers markets. Support regional farmers who sell seasonal food.

11.3 Become a community-supported agriculture member. For a CSA near you, check out the LocalHarvest website.[37]

11.4 Go gleaning. Check out the Falling Fruit interactive map with over 1.4 million locations to forage edibles within our cities.[38] No carbon-spewing tractors required.

11.5 Buy bulk ingredients and store them in bulk containers. Save packaging.

11.6 Buy Demeter Certified Biodynamic products. They go beyond organic sustainability standards. Look for them at Whole Foods markets.[39]

11.7 Serve vegetables, fruit, and nuts. Serving earth-friendly food creates Handprints. You influence the eating habits of others.

11.8 Serve organic food. Save the soil. Influence farming and marketing practices.

11.9 Serve perennial food such as berries from bushes, asparagus, rhubarb, kale, watercress, and horseradish.[40] Help build markets for foods that sequester carbon.

11.10 Serve raw food and fermented food. Save cooking energy.

11.11 Explore plant-based alternatives to animal products. Try egg-free mayo. I like Tofurky Italian sausage. Check out the free food guide from Compassion in World Farming.[41]

11.12 Create earth-friendly recipes for all of the above. Calling all chefs!

11.13 Let go of beef. And check out the other recommendations on the Eat Low Carbon website.[42]

11.14 Teach others to serve earth-friendly food. Suggest that your local store or community center host eco-cooking demonstrations. Make it patriotic.

11.15 Write online reviews praising stores and restaurants that meet an eco-need.

IN YOUR JOURNAL

What steps will *you* take to create culinary Handprints?

V

SHELTER, MOTION,
AND ENERGY HANDPRINTS

12

LIVE AND WORK
IN MODEST ELEGANCE

When our youngest son moved out, my wife, Willow, announced that she wanted to move to a small home. So, she found a 518-square-foot house on a half-sized city lot. Okay. We bought it—and instantly cut our house payments in half. Our heating bills were cut in half as well. I spent hours culling my files to fit into this unassuming little dwelling. Since it was long and narrow, I called it a land yacht. The idea of living as if on a boat helped me to experience our move as an adventure. A life-changing payoff arrived four months later: lower monthly payments allowed me to accept a voluntary early retirement.

We need to find modestly elegant ways to live in the world. When we do, a Handprint is created.

SMALL HOUSES

We have lots of room for downsizing Handprints. The average US family size *decreased* from 3.67 people in 1948 to 2.55 people in 2012. In 1978, the average US house measured 1,525 square feet; by 2013, it *grew* to 2,598 square feet.[1] Right-sizing our home creates sustainability by using less energy, reducing the impact on the land, and freeing up money for worthwhile causes. You do not have to go "tiny," but keep in mind that small can be beautiful.

TIPS FOR LIVING WELL IN A SMALL HOUSE

- **Ask, "Do I/we still need this?"** We let go of our clothes dryer, our big refrigerator, a desk, and two bedrooms that the boys, who moved out, no longer needed. We kept the clothes washer.
- **Ask, "Does this spark joy?"** That useful question comes from *The Life-Changing Magic of Tidying Up* by Marie Kondo. I applied the test to whole drawers of my file cabinets.
- **Use vertical space.** Hang utensils. Install shelves and lofts.
- **Use folding furniture** for maximum versatility of space. Folding chairs. Folding tables. A futon that folds out into a bed. Our rocking chair, clothes drying rack, and bookshelves all fold up.
- **Use an outbuilding** if your preferred lifestyle is not exactly spartan. My boxes full of memories do not need to be heated.
- **Entertain in someone else's house.** If you bring dinner, you will always be welcome.
- **Put the house on wheels.** You do not have to buy the land on which your house sits. Being able to move can be a plus.

The Small House Society lists eighty US small home designers and builders. Literally a cottage industry! Company names convey the spirit of a small revolution: Alchemy Architects (Minnesota), BC Mountain Homes (British Columbia), Bear Creek Carpentry (New York), Creative Cottages (Maine), Non-Compliant Design (Washington), Little House on the Trailer (California), Turtle Back Nomadic Yurts (Colorado), Tortoise Shell Homes (California), and Texas Tiny Homes. In San Francisco, New Avenue Homes specializes in small plot development. First Day Cottage in New Hampshire specializes in small home kits. DIY-Homebuilding in Wisconsin will support you in designing and building your own small home—anywhere.[2]

Tiny houses do not have to be *your* permanent residence. They can be rented out or sold into a housing market hungry for modestly priced dwellings.

ECO-REMODELING

Recycling a house is certainly in the realm of sustainability. Any change that keeps on working no matter who occupies the house is certainly a Handprint. A quiet revolution awaits the eco-connoisseur of previously owned homes.

After living in a small house for three years, Willow and I moved into a fixer-upper. Over the years, we upcycled the house with many of the options described below.

A MENU OF ECO-FRIENDLY HOME REMODELING OPTIONS

- **Efficient windows.** Windows have evolved from single glazed to double glazed to windows with heat-trapping coatings to windows independently tested and rated by the National Fenestration and Rating Council.[a] That experience is now built into every Energy Star label on Milgard and Anderson windows, doors, and skylights.[b]
- **Tankless (on-demand) water heaters** eliminate the standby heat losses from tanks. Save between 8 and 50 percent of your water heating cost. Families with low water usage save more. The highest savings come from dedicated units that serve the bathroom, kitchen, and/or laundry. Our wall-mounted Rinnai on-demand heater frees up valuable floor space.[c]
- **Front-loading clothes washers** use five gallons less water per load and less detergent than top loaders—at half the en-

---

[a] "Questions about Buying New Windows Doors or Skylights," National Fenestration Rating Council, accessed January 5, 2016, http://www.nfrc.org/WindowRatings/.

[b] "Energy Star," Milgard Manufacturing, accessed February 5, 2016, http://www.milgard.com/learn/energy-efficiency/energy-star. "Energy Star," Anderson Windows and Doors, accessed February 5, 2016, https://www.andersenwindows.com/planning/articles/energy-star/.

[c] "Tankless or Demand-Type Water Heaters," Energy Saver, Office of Energy Efficiency and Renewable Energy, U.S. Department of Energy, accessed November 8, 2019, https://www.energy.gov/energysaver/heat-and-cool/water-heating/tankless-or-demand-type-water-heaters.

- ergy. Front-loaders clean better and wring clothes out better. Manufacturers have debugged the reliability issues.[d]
- **Clothes drying racks.** Using a foldable or drop-down clothes-drying rack shaves more than half a ton off a family's carbon footprint every year.[e] The Japanese hang clothes, even in luxury apartments.[f]
- **Automatic dishwasher.** Automatic dishwashers are more efficient than all but the most frugal handwashers.[g] Make sure the model you install qualifies for an Energy Star label.
- **Dual flush toilets and low-flow showerheads** are now readily available. Insist on them.
- **LED lighting** is now inexpensively available for every screw-in fixture. Seek advice from the hardware store or an electrician when replacing long fluorescent tubes.
- **Separately switched lights** eliminate waste as well.
- **Motion sensors** are an energy-saving standard these days.[h]
- **Low-VOC paints** make a home feel new without the release of hydrocarbons into the air. We used Miller Acro Pure paint. Benjamin Moore, Sherwin Williams, and Behr also offer low-VOC paint.
- **Recycling latex paint** has evolved into a profitable enterprise with remarkably consistent color choices. Check out Amazon Select Paints.[i]
- **Buy used building products.** Our eco-remodel reused bathroom and kitchen cabinets, a sink, and two interior doors, mostly from Portland, Oregon's, Rebuilding Center.

[d] Daniel Wroclawski, "The Great Washer Debate: Are Front-Loaders Really Better?" *USA Today*, October 13, 2014, http://www.usatoday.com/story/tech/2014/10/13/the-great-washer-debate-are-front-loaders-really-better/17204535/.

[e] Author's calculation of savings from eliminating power clothes dryer: 5 loads/week × 50 weeks/year × 3.3 kWh/load × 1.5 pounds of CO_2/KWh × ton/2,000 pounds = 0.62 tons/year. Rack drying would also increase home heating energy use in winter, which may be more than offset by less need for air-conditioning in summer, depending on the local climate. Family size will affect number of loads per week.

[f] "Clothes Dryer Energy Use," Saving Electricity, Michael Bluejay, accessed March 6, 2016, http://michaelbluejay.com/electricity/dryers.html.

[g] Mitchell Parker, "Dishwashers vs. Hand-Washing Debate Finally Solved—Sort Of," *Houzz*, December 22, 2015, http://www.houzz.com/ideabooks/58245033/list/dishwasher-vs-hand-washing-debate-finally-solved-sort-of.

[h] Story told to Jon Biemer by Ken Keating during an interview in 2008.

[i] "Amazon Select Paints," Amazon Paints, accessed November 8, 2019, http://www.amazonpaint.com.

- **Require Forest Stewardship Council certification** when building with new wood. "FSC's forest management standards expand protection of water quality, prohibit harvest of rare old-growth forest, prevent loss of natural forest cover and prohibit highly hazardous chemicals, which are all unique aspects of the [FSC] system."[j]
- **Light tunnels, solar collectors, graywater recovery, and rain storage** have all evolved since the days of hippy experimentation.
- **Heat pumps** now have a long track record of reliably eliminating fossil emissions associated with oil and natural gas heating—*if* you buy electric power that is not generated by fossil fuels.

---

[j] "Advantages of FSC," Forest Stewardship Council, accessed August 16, 2019, https://us.fsc.org/en-us/what-we-do/advantages-of-fsc.

When the roof needed replacing, we asked for extra insulation. We found that installing an on-demand tankless water heater increased the value of the house by allowing us to finish the space previously occupied by the old water heater tank. We had to insist that the installer of our ASKO energy-efficient dishwasher hook it up to the *cold*-water line. It internally heats only the little bit of water it needs. Likewise, our ASKO clothes washer heats its own water as needed. (Most of the time we wash our clothes in cold water.) To raise awareness about our eco-adventure without bragging, Willow and I posted signs asking "What's green about this room?" and listing eco-features throughout the house.

MULTIFAMILY HOUSING

The US Energy Information Administration reports that *multifamily units use half as much energy as single-family homes.*[3] Put differently, the mass migration of people from rural areas to the cities is good news, environmentally speaking.

Condominiums, apartments, and multiplexes draw upon many of the efficiency techniques used for residences, such as installing Energy Star appliances. In addition, heating, ventilating, and air-conditioning systems can transfer heat from one part of the building to another. Centralized laundry facilities can be offered. Exterior walls can be used as solar collectors. Most significant, common walls between apartments reduce each family's winter heat loss to the outdoors.

The higher density of people in apartments improves the practicality of public transportation systems and overall public infrastructure. On a per capita basis, city dwellers use less energy than small-town people or rural people (in that order).[4] City apartment dwellers can take pride in their modest Footprint. For instance, fewer than half of New York City households even own a car. Over 80 percent of New York City employees travel to work on foot, bike, or mass transit. If New York City were a state, it would be fifty-first in per capita energy use![5]

Whether you live in an apartment in Chicago or Beijing, you have at least one reason to take pride in your dwelling.

LEADERSHIP IN ENERGY AND ENVIRONMENTAL DESIGN

That apartment building itself may reflect the profound impact of an extended family of eco-friendly professionals organized by the US Green Building Council. In August 1998, the council launched the Leadership in Energy and Environmental Design (LEED) certification system.[6]

A building earns points based on its eco-features—the more, the better. It can simply be LEED certified at a minimum level. Higher scores earn ratings of Silver, Gold, and, for extraordinary buildings, Platinum.

Initially, architects, builders, and owners faced significant challenges in meeting LEED standards. Standard practices needed to change. Sustainable product suppliers were hard to find. The documentation required for independent verification was onerous. But, one by one, where civic support, professional expertise, and owner flexibility converged, LEED buildings were built. I was privileged to serve on a review committee of Portland, Oregon's, Green Investment Fund, which sponsored several LEED demonstration buildings.

Gradually the building environment changed. Hard-to-find eco-technologies became more available, like windows that can be switched to let in more or less light. Seattle started requiring LEED certification for its municipal buildings.[7] Design teams started using integrated project delivery principles, which treated the building as a whole system rather than as a series of features imposed sequentially by the owner, the architect, the structural engineer, the mechanical engineer (space conditioning), and the electrical engineer (lighting and computers).

And LEED itself evolved, fueled by an expanding circle of LEED-certified professionals. Features that could earn points expanded, such as hardwood certification for sustainable sourcing and surface water management (landscaping and parking lots). LEED certifies homes now. Schools receive points for a reduced parking Footprint and joint use facilities. Hotel (hospitality) developers claim rainwater management and green cleaning policy. Points for retail stores come from storage and collection of recyclables and daylighting. Warehouses and distribution centers are rewarded for sensitive land protection and green vehicles. Measures for LEED-certified neighborhoods now include tree-lined streets, transit facilities, local food production, and redevelopment of brownfields, such as old gas stations. As professionals gained experience, LEED standards changed to allow creative (in addition to prescriptive) approaches to meeting challenges such as water conservation.[8]

Over time, the voluntary LEED standards have made it easier to include sustainability features in state and local building codes. In 2018, the US Green Building Council cosponsored the International Green Construction Code, which offers jurisdictions around the world a model for earth-friendly building practice.[9]

In 2018, Curitiba, Brazil's Petinelli consulting headquarters became the first building to achieve LEED Zero Certification.[10] To achieve that distinction, it had to demonstrate zero net carbon emission, energy use, and water use over a year. It needed to meet LEED's Zero Waste certification criteria as well.[11]

Fundamentally, LEED is greening mainstream building practices around the world. There are more than ninety-six thousand LEED buildings in 167 countries and territories.[12] Canada, China, India, and Brazil have the most LEED buildings outside the United States. Half of the 200,000 certified LEED professionals live and work outside the United States.[13] For them, creating Handprints is big business.

HANDPRINT OPPORTUNITIES

12.1 Install LED lights. LED lights are available for nearly all fixtures.

12.2 Choose Energy Star–labeled appliances, lighting, windows, computers, and new homes.

12.3 Use clothes drying racks and hanging bamboo rods instead of electric or gas dryers.

12.4 Use a food box on a north-facing porch instead of a refrigerator. Milk lasts two to four days. The grocery store is our community refrigerator. Local restaurants serve meat if I want it.

12.5 Buy used house parts. Check out one of the 900 ReStores, each of which supports Habitat for Humanity.[14]

12.6 Post "What's green about this room?" signs around the house. Visitors notice.

12.7 Eco-retrofit your home. Consider possibilities listed in this chapter. Help eco-industries in the process.

12.8 Move to a more earth-friendly home. Consider these attributes:

- Small size. Cash out those empty bedrooms.
- Shared walls. Multifamily housing requires far less energy per person.
- Near public transportation. Bus lines have eco-value.
- Near employment. Or telecommute.
- In the city. Greater density means a lower per capita impact on nature, access to public transportation, and availability of multi-family housing.[15]
- Solar access. A south-facing roof in the Northern Hemisphere. Beware of shading structures. (See Handprint Opportunities 12.9 and 12.10.)
- LEED certification.
- Eco-remodel potential.

12.9 Salvage tear-down buildings. Up to 80 percent of an old building can be recovered, donated, and claimed as a tax deduction.[16] Check out The ReUse People, with offices around the country. Independent deconstruction companies can be found in Texas, Montana, Virginia, and Oregon.[17] Some ReStores (Handprint Opportunity 12.5) offer nonstructural deconstruction services as well.

12.10 Live in a solar-friendly house. In temperate climates in the Northern Hemisphere, houses with an east-west orientation and a predominance of south-facing windows use less energy. Use overhangs to shade the summer sun and welcome the low winter sun. Plant solar-friendly trees that lose their leaves early in the fall.

12.11 Promote earth-friendly energy in your community. The North Carolina Clean Energy Technology Center maintains DE-

SIRE, a free online database of renewable-energy and energy-efficiency incentives and policies. Check out your location.[18] Boulder, Colorado, and Ashland, Oregon, have solar access ordinances that restrict a neighbor's remodel from shading your solar collector.[19] Kansas City, Missouri, codified a process to negotiate an easement to access solar and geothermal (ground-source) energy.[20]

IN YOUR JOURNAL

How will *you* make your home and/or place of work more elegantly sustainable?

13

MOVE WITH GRACEFUL ECONOMY

How we get around matters. Handprint thinking suggests we study changes in lifestyle, user-friendly technology, and transportation planning. Still, we evolved to walk. Why not begin our exploration of sustainable mobility with our own two feet?

WALKING

Over the course of a day, an active person takes around five thousand steps. Over a typical lifetime, a person can walk 100,000 miles—four times around the earth.[1] The human foot has twenty-six weight-bearing bones plus some two dozen small sesamoid bones.[2] Sesamoid bones are embedded in tendons either between bones or alongside larger bones. They work like pulleys to reduce friction, allow joint movement, and increase muscle effectiveness.[3] Amazingly, this complicated system, wrapped in fascia and skin, is quite durable.

Walking is nonpolluting. The exercise helps us stay healthy. Walking allows us to truly appreciate the landscape. We are seen while walking. We deepen relationships with each other and with our community while walking. And walking can get us where we are going. It is even arguable that our ability to walk is part of what makes us distinctly human.

According to City Lab, "Less than 3 percent (2.7 percent) of Americans walk to work. But more than 5 percent of workers do in New York City (5.9 percent), Honolulu (6.5 percent), and Boston (5.2 percent). An even larger share walks in smaller metros and college towns including Flagstaff, Arizona (9.7 percent) [and] Iowa City, Iowa (8.7 percent)."[4]

However, these data do not take into account those who raise children and take care of the elderly—predominantly women. In *Invisible Women*, Caroline Criado-Perez reports that male-dominated transportation planning processes often marginalize (and don't even measure) nonlinear, multipurpose trips.[5] "Trip chaining" is typical of women raising children and caring for the elderly. Citing European data, Perez notes that women are the majority users of sidewalks and buses. Reinforcing this conclusion, in the United States, 55 and 56 percent of bus riders and train riders, respectively, are women.[6] Put simply, *removing gender bias, however unintentional, makes a case for environmentally preferable transportation.*

THE BICYCLE

Bicycles are five times as efficient as walking.[7] We can thank Handprint innovations like equal-sized wheels, air-filled rubber tires, and chain drives for the most efficient form of overland transportation.[8] Considering caloric burn, bicycles can get the equivalent of 900 miles a gallon (383 kilometers per liter) of gasoline. That is thirty times as efficient as a typical car.[9]

The percentage of the US population who biked to work in 2018 was about 0.6 percent, a number that has stayed relatively constant in recent years.[10] Hopefully, this number will increase as more and more cities complete plans to reduce inner-city traffic congestion—like Budapest, Salt Lake City, and Portland, Oregon.[11]

Electric bicycles, available since the 1990s, offer many of the efficiencies of a traditional bicycle without the exercise regimen. The Espin eBike retails for under $2,000, a fraction of the cost of a serviceable box on four wheels.[12] E-bikes can go around 800 miles on the electrical equivalent of a gallon of gasoline (340 kilometers per liter).[13]

Electric scooters that can be rented for rides from random street locations are making a debut in numerous cities. Complaints of disruption and unruly riders notwithstanding, their availability also improves the prognosis for our environment.

THE ELECTRIC CAR

The electric car is a disruptive technology.

Several converging factors spell doom for the gasoline-fueled piston engine car. First, the electric motor is three times as efficient as a piston engine.[14] Second, locating the motor at the wheel eliminates the weight of the drive train. Third, recharging the battery while braking saves more energy yet. Fourth, electric cars require less maintenance and go farther before wearing out.[15] Fifth, the energy storage capacity of electric batteries has steadily improved. Sixth, lightweight carbon fiber, once the exclusive domain of military aircraft, can now be molded into car bodies.[16] Seventh, the price of solar and wind energy has dropped such that cars on the road do not justify more coal or nuclear power plants.

We can thank MacArthur Genius Grant winner Amory Lovins for putting several of these pieces together in the Hypercar concept in 1991—then releasing the fundamental principles into the public domain so no one manufacturer could appropriate or sequester them.[17]

Nissan first offered the all-electric LEAF (Leading Environmentally-friendly Affordable Family car) in 2010. Over the decade the LEAF has been on the market, Nissan has roughly doubled its range—up to 226 miles (364 kilometers) at the equivalent of 108 miles per gallon (2.2 liters per 100 kilometers) of gasoline.[18]

Tesla Motors initially focused on the sports car market—high performance and expensive. Silicon Valley founders Martin Eberhard and Marc Tarpenning boldly (foolishly?) built their all-electric vehicle outside the factories and distribution networks of major automakers.[19] They presold classy Teslas to comparatively wealthy risk takers ("innovators" in marketing terminology) who would wait months, even years, for delivery. The Tesla's carbon-fiber body required costly and time-consuming manufacturing innovation.[20] Built into the frame of every Tesla is a lithium ion battery. A gigantic lithium ion battery factory in Nevada complements Tesla's intention to "accelerate the world's transition to sustainable transport."[21]

The all-electric Chevrolet Bolt has been on the market since 2016. It offers a combined mileage of 119 miles per gallon (2.0 liters per 100 kilometers).[22] And it takes a family of five 238 miles (383 kilometers) between charges.[23] Chevrolet succeeds in offering those features at a competive cost.

Charging infrastructure must keep pace with electric vehicles on the road. There are at least twenty-two thousand electric vehicle (EV) charging stations in the United States.[24] PlugShare now has a map-accessible database.[25] Google can now show you the real-time availability of the nearest charging station.[26] That is, in addition to your home charger.

We have not phased out the internal combustion engine yet. *Inside EV* reports that 2018 EV sales were 360,000 in the United States and over

2 million worldwide.[27] This compares to 17 million total US vehicle sales in 2018.[28]

We have a long way to go, but purchasers of hybrids and then EVs have helped us get this far. As of 2019, at least thirty-nine manufacturers offer EVs for sale. Ten of those offer ranges greater than 200 miles (320 kilometers).[29]

A LOW-CAR LIFESTYLE

The environment reaps an additional reward if we actually *avoid owning a car*. The embodied carbon Footprint of making a car can be eliminated. The emissions associated with making a car include building and operating the factories that make it, harvesting and mining the raw materials (rubber and iron), shipping the finished vehicles, and maintaining a dealer network. Numbers vary, but investigative reporters for *The Guardian* newspaper assert, "The embodied emissions of a car typically rival the exhaust pipe emissions over its entire lifetime."[30] That embodied Footprint remains even when we move to an electric car.

Forgoing car ownership also saves money on insurance, maintenance, and depreciation. For gasoline-powered vehicles, the American Automobile Association tells us that the average car costs its owner $8,698 per year.[31] Even if an electric vehicle halves that cost, owning a car may cost more than it is worth.

For these reasons, Willow and I lived car-free for thirteen years.

JON'S TIPS ON LIVING WITHOUT A CAR

Embrace the opportunity presented by a new home or job. Every time an old vehicle wears out, question whether it needs to be replaced.

Make mobility an art. My wife and I rode buses to see a total eclipse—our eco-eclipse. A friend linked together five bus *systems* to hike a fifty-mile (eighty-kilometer) section of the Pacific Crest Trail. Use transit time to read or cull surplus photos from your cell phone.

Use delivery services especially for bulky items like toilet paper. Home delivery is a new norm in the wake of COVID-19.

Use shared transportation as needed. We call a cab to go to the emergency room. We sometimes rent gently used cars from Portland's local Crown Auto Rental for vacations.
Treat your friends well. Feed them during and between rides. Be generous with gas money; give extra for "wear and tear." Many drivers will refuse payment to support your living without a car.

COVID-19 will accelerate three trends that already accommodate living without a car. First, the number of people working at home was already increasing. Second, the economy of remote meetings, via Zoom and other platforms, was already curtailing work-related travel. Third, Amazon had already established a viable home-delivery market. Now other retailers have followed suit. With our collective experience with "stay home; stay safe," these trends have been supercharged.[32]

The availability of shared cars also facilitates living without one in the driveway or garage. At one time, hitchhiking or taking a taxi were the only ways of getting across town without owning a car. Now Uber and Lyft offer owner-driven rides to wherever you want to go. Waze offers apps that will help you find rideshare partners who are already going your way—in real time.[33] Zipcar and Enterprise allow you to rent a car by the hour.[34] Turo offers a peer-to-peer platform in which you can rent directly from other car owners, much like Airbnb does for places to stay.[35]

Political will is now such that in some cities, pedestrians and bicycles are winning priority over cars. Cities that restrict cars in the city centers include New York City, San Francisco, Mexico City, Brussels, Copenhagen, Oslo, and Madrid. Hamburg, Germany, is creating a green network of car-free zones that, by 2025, will comprise 40 percent of the city. Chengdu, China, is dedicating half its streets to pedestrians and bicycles.[36]

Of course, excellent public transportation complements any car-free lifestyle.

URBAN MASS TRANSIT

The International Association of Public Transport has 1,800 members from one hundred countries.[37] New York City's subway system is the largest in the world—232 miles of routes and 421 stations. The New York City subway, at 1.75 billion rides a year, 5.6 million rides a day, is number six in ridership—after Beijing, Seoul, Moscow, Tokyo, and Guangzhou.[38] At least 160 cities have operating rapid transit systems, with another 25 under construction. This is large-scale sustainability!

Jaime Lerner, an architect in Curitiba, Brazil, championed the world's first bus rapid transit (BRT) system in 1974.[39] The BRT includes dedicated bus lanes, off-vehicle ticketing, prioritized signaling to avoid stops for traffic, and station platforms level with the bus floor. BRTs come in at less than half the cost of light rail systems.[40]

Five miles of the ride to Pittsburgh airport is on its BRT, dedicated in 1979. After sixteen years, Quito, Ecuador, followed suit in 1995, then Bogotá, Colombia (2000), and Jakarta, Indonesia (2004). BRTs now provide over thirty-three million rides a day in 170 cities![41]

Back on September 12, 2015, I joined thousands of others to celebrate the opening of Portland's Orange Line. The Orange Line adds 7.3 miles to our light rail system, which opened its first line in 1986. The iconic heart of the new line is a bridge over the Willamette River called Tilikum Crossing, which, roughly translated from the Chinook language, means Bridge of the People. You can walk or bike across it; you can ride the streetcar, light rail, or bus across it; but you cannot drive across it. The Orange Line took four years and $1.49 billion to build, but you can be sure that this carefully designed piece of infrastructure will serve our grandchildren.[42]

INTERCITY RAIL

There are at least two reasons why a train is more efficient than a car, truck, or bus. The friction on steel rails is less than rubber on pavement, and the face of the vehicle, which pushes air aside, is small compared to the number of people and amount of materials being moved.[43]

"We may ask you to move so families can sit together," says the conductor of the *Empire Builder*. The early part of my trip from Portland, Oregon, to Shelby, Montana, follows the Columbia River. The Wi-Fi and line power plug-in turned my seat into a mini office—when I chose to take my eyes off the scenery. The next day, we passed the south edge of Glacier National Park, where a volunteer entertained us with stories of the towns along the way. Not once did I hear a complaint from my fellow riders.

In 1992, Congress passed the Amtrak Authorization and Development Act, which resulted in the Acela Express service—a high-speed rail service between Boston; New York City; Philadelphia; Washington, DC; and Baltimore.[44] The legislation came with a goal: New York to Boston in three hours. Engineering problems embarrassed the railroad for years. With a top speed limited to 150 miles per hour (241 kilometers per hour), the best New York City–to–Boston time was 3.5 hours. Even so, incremental improvements (think Handprints) in the service made the Acela Express, Amtrak's "Northeast Corridor," both popular and profitable. Eighty percent of the seats are sold in some segments. The Acela Express netted an operating profit of $334 million in 2019.[45]

A Stanford University study shows that high-speed rail can replace airlines when cities are separated by, say, two hundred to three hundred miles (three hundred to five hundred kilometers), or up to a 4.5-hour train ride. Trains deliver people to city centers (as opposed to the outskirts, where most airports are located), are more economical, and have less need for security checks. So there are no more commercial plane flights between Paris and Brussels.[46] That is good news for the environment. Jets are less efficient, and water vapor released into the upper atmosphere has an added greenhouse effect (referred to as the "radiation forcing index" or "aviation impact factor").[47]

High-speed rail is now a mature technology. In October 1964, Japan inaugurated the Tokaido Shinkansen, known in the West as the Bullet Train, between Tokyo and Osaka. It traveled 320 miles (515 kilometers) at an average speed of 101 miles per hour (163 kilometers per hour). France, Germany, and Spain followed suit in 1981, 1991, and 1992, respectively— each with its own manufacturing capacity. Belgium, the United Kingdom, South Korea, Taiwan, the Netherlands, and Turkey have since brought online their versions of high-speed rail. Russia, Morocco, Saudi Arabia, and Eastern Europe are getting serious about high-speed rail too.[48] China unleashed its high-speed rail potential in 2008, creating its own cars and opening the first stage of a huge national system. China now has over 18,000 miles (29,000 kilometers) of high-speed rail installed, two-thirds of the world's total.[49]

The US High Speed Rail Association has long advocated going beyond the Northeast Corridor. In 2018, All Aboard Florida put the first sixty-five-mile stretch of its high-speed rail service from Miami to West Palm Beach. The Brightline, as it is called, is privately funded. Future phases, stretching to Tampa, will add another 240 miles to the system. Florida Department of Transportation's decision in the 1990s to reserve the whole corridor for high-speed trains made this project viable.[50] Texas Central Railroad, also privately financed, is constructing high-speed rail between Dallas and Houston that will be linked with the Amtrak interstate rail network.[51] Go, teams!

In January 2015, California broke ground on its high-speed rail line from Los Angeles to San Francisco.[52] Voters seeded the project with a $10 billion bond in 2008 and an innovative carbon cap-and-trade law provided additional funding. However, high cost (the original concept envisioned tunneling) and reduced economic activity due to COVID-19 threatens the viability of that project. I love the idea of northern and southern California being connected by high-speed rail. Still, it is not a panacea any more than bicycles and electric cars.[53]

THE NEXT GENERATIONS

Millennials (born between 1981 and 1996) and generation Z (born after 1997) both give us plenty of reason for hope. The 2017 National Highway Transportation Study states, "The trends over the last two decades clearly indicate that . . . overall trip-making are declining, with larger declines noted for younger people."[54] Likewise, a 2018 survey by Cox Automotive finds that owning a vehicle is not important to 45 percent of millennials and to 55 percent of generation Z but that transportation alternatives *are* important.[55]

Why?

These generational eco-friendly trends do not depend on heightened environmental values. Young people these days express identity more through social media than by having a car. Increased student debt burdens make it harder to afford a car. Most states now have a graduated approach to earning a driver's license (learner's permit, limited privileges, then full privileges). This process discourages some from getting a driver's license at all. Concurrently, more young people live in cities, where efficient transportation options (bus, light rail, on-call rides, shared rides, bicycles and scooters) are more available. One other factor: young people in the United States are having children later and having fewer of them. Thus, their need for family-sized people haulers is less.[56]

The impact of this admixture of demographics and Handprints is profound. A 2018 Massachusetts Institute of Technology study observes that "on average Millennials own approximately 0.4 fewer vehicles per household than the average Baby Boomer." A 2018 survey by marketing consultant Allison+Partners, reports that "nearly 70 percent of Gen Z respondents do not have their driver's license and 30 percent of those who do not currently possess their driver's license have no intention or desire to get one."[57] This disenchantment with car ownership is relevant. The millennial generation and generation Z comprise more than 60 percent of the world's population.[58] Add to that the conclusion of a 2018 report by professors at the University of California Los Angeles: convenient access to a car may be the single biggest deterrent to transit ridership.[59] Put more positively, the next generations are ready to embrace elegantly efficient mobility.

HANDPRINT OPPORTUNITIES

13.1 Walk or ride a bike whenever possible. To work. To the store. To have fun—on the way and when you get there. The more people who eco-travel, the more of an eco-travel culture we become.

13.2 Arrange online meetings. Think Free Conference Call, Skype, and Zoom.

13.3 Carpool. Give someone who came by bus a ride home. Set up a carpooling program. Check out RideAmigos.[60]

13.4 Trade in your gasoline car for an electric car. Help phase out the gas station.

13.5 Live without owning a car. Save the embodied carbon emissions of a driveway ornament.

13.6 Take children on the bus so they will get used to it. Urge them to use public transit.

13.7 Drive a bus or a train so others can stay off the road.

13.8 Participate in public transportation planning and development processes:
- Support sidewalks and bike lanes.
- Ensure adequate electric vehicle charging capacity.
- Support car-free urban zones and surcharges during peak traffic periods (congestion pricing).
- Advocate for public transit.

IN YOUR JOURNAL

What is *your* personal vision for moving with graceful economy?

14

INSIST ON
EARTH-FRIENDLY ENERGY

Dear Jon," I imagine my grandchild writing to me in a decade or two. "Thank you for your two years' service with the Solar Rating and Certification Corporation (now part of the International Code Council) to improve the quality of solar water heating collectors. You were there," he or she would continue, "to arrange utility funding of efficient window research in the Reagan administration. And thank you for helping plan the world's first industrial energy-efficiency conference. Wow! You helped improve the efficiency of water and sewage treatment plants for Kalispell, Montana; Jackson, Wyoming; and Richland, Washington." Alas, this letter is a fantasy. Even my bosses at Bonneville Power Administration would not remember these particular technical, policy, and political steps to scale up renewable energy and energy efficiency. They are Handprints nonetheless.

Renewable energy has rightly gained an increasingly respected place in our planetary energy portfolio. That said, the most valuable form of renewable energy is the energy that you no longer even want to use.

ENERGY EFFICIENCY

Improving energy efficiency creates a Handprint by saving energy even when those who put it in place move on.

Back in the 1980s, energy conservation professionals working for progressive utilities—myself included—adopted a mantra: save energy first; then meet the remaining needs with renewable energy. Utilities and their contractors plugged air leaks and added insulation ("weatherized") to hundreds of thousands of houses based on this philosophy, often offering financial incentives to do so. Chapters 12 and 13 make the cases for energy efficiency in buildings and transportation, respectively.

My main objective in moving from a consulting engineering firm to an electric utility was to be part of a program that averted the need to build big power plants. In that, I was successful. A certificate on my wall attests to the fact that my colleagues in the Pacific Northwest reduced energy demand enough to avoid building two large centralized fossil or nuclear power plants.

However, in 1996, electric utility deregulation disrupted utility energy-efficiency programs as we knew them. Electric utilities were no longer responsible for meeting the growth of demand for electricity in their area. Consumers could buy from other utilities.

That crisis among efficiency-minded people gave birth to a fresh line of thinking. Instead of hoping the market would automatically deliver cheap, reliable, customer-friendly energy efficiency, let's try intervening in the market. The new mini industry was called "market transformation."

The purpose of market transformation is to disrupt market behaviors that tend to perpetuate the consumption of a lot of electricity. Such behaviors include people's and businesses' natural tendency to stay with outdated technology as well as marketing campaigns that have little to do with efficiency. The tools of market transformation came to include demonstrating new technologies, training equipment operators, coordinating energy-efficient building code development, and encouraging manufacturers to respond to consumer concerns.

The Northwest Energy Efficiency Alliance carries on this market transformation vision. It is funded by Bonneville Power Administration and fifteen other electric and natural gas utilities in the Pacific Northwest, and it names 140 organizations as constituents, including state energy offices and environmental nonprofits.[1] Northwest Energy Efficiency Alliance projects in the residential, commercial, institutional, and industrial sectors have had a national impact.

Nationally, the Institute for Market Transformation (IMT) also used the principles of market transformation to help write revised building codes for Washington, DC, (2006) and to implement New York City's Greener, Greater Buildings Plan (2009). IMT notes that 80 percent of the buildings that will be standing in 2030 already exist today. According to their website, "IMT crafts and activates local networks to deploy efficiency faster and deeper in all stages of building use, from financing, design, and construction, through occupancy, tenant turnover, renovation, and disposition."[2] Through the City Energy Project, IMT and the Natural Resources

SELECTED INITIATIVES OF THE NORTHWEST ENERGY EFFICIENCY ALLIANCE (1997 TO PRESENT)

- **Building codes.** For decades, the Northwest Energy Efficiency Alliance (NEEA) facilitated the adoption of residential and commercial energy-efficiency building codes in the Pacific Northwest.
- **Washing machines.** NEEA promoted the sales of energy-efficient horizontal axis washing machines, which use less water and wring clothes better (for quicker drying) than machines with a vertical agitator. Horizontal axis washers are now common.
- **Heat pump water heaters.** A field validation study of the remote-controlled heat pump water heaters demonstrated that utilities can reduce the evening water heating electric demand by 90 percent—and save energy. This study helps document performance claims and inform distributed-energy operation protocols.
- **Windows.** NEEA worked with manufactures and distributers to include efficient Energy Star windows in new homes before selling them to consumers.
- **Building operation.** Improper installation and operation of building heating and air-conditioning systems wastes energy. NEEA supported efforts to "commission" buildings after they are built and certify building operators.
- **Hospitals.** A ten-year initiative to demonstrate strategic energy management in hospitals showed that energy used in hospitals could be a managed cost (reduced by up to 20 percent) rather than a fixed cost. Other hospitals in the Pacific Northwest and elsewhere have adopted strategic energy management protocols.
- **Motors.** "Motors use about half of all U.S. electricity." Partnering with a national effort, NEEA helped industrial motor users incorporate life cycle analysis into their motor purchase and rewinding decisions.
- **Industries.** NEEA worked with industry associations to promote energy efficiency in member pulp and paper mills and food processing facilities.

Source: Northwest Energy Efficiency Alliance website.

Defense Council promoted large building energy efficiency in twenty cities (2013–2018), including Des Moines, Iowa; New Orleans, Louisiana; Providence, Rhode Island; and Reno, Nevada.[3]

WIND

The wind contributes nearly 6 percent of the world's electric power.[4] That is a big green number. Windmills are ubiquitous in areas ranging from Texas to Oregon to Mongolia. Wind power and its cousin, solar power, are displacing fossil fuels. We can and should celebrate that extraordinary progress.

But where did such a monumental Handprint come from? It came from the Persians, who built the first grain-grinding windmills around 800 CE. It came from English commoners, who learned how to turn their mills into the wind. It came from the Dutch, who learned to pump water with wind. It came from Poul la Cour, who guided the construction of forty demonstration windmills between 1891 and 1907—with the support of the Danish government. It came from wind tunnel studies done in 1918 by Albert G. von Baumhauer, which established the primacy of the propeller-shaped wind "sails" we know today. It came from the Jacob brothers, who designed and sold twenty thousand electricity-generating windmills, many for rural farmers, from 1932 into the 1950s. It came from Palmer Putnam, in the 1940s, who carefully documented the success and eventual failure of his extra-large windmill on Grandpa Knob in Vermont. We have a lot to thank our ancestors for.

The Public Utility Regulatory Policies Act of 1978 insisted that companies other than established utilities could generate and sell power through existing transmission lines. This set the stage to demonstrate twelve thousand windmills in the California passes from 1983 to 1986. But there were embarrassing and expensive reliability problems. You could see from nearby freeways. Many windmills on those California passes stood still. Most of the windmill manufacturers went broke.

But not the Danish participants. Their eighty years of experience spelled durability. They used the large-scale California demonstration to launch a worldwide wind revolution. Vestas, a Danish company, is (as of 2020) the world's leading installer of windmills.[5] Besides, we collectively learned from the failures. Generators are now placed at the hub of the rotors rather than on the ground to avoid mechanical vulnerabilities arising from mechanically transmitting rotating energy. Three blades on rotors proved more cost-effective than two blades. And cylindrical columnar towers replaced lattice towers and their cross braces, which fatally attracted birds.

Policy innovations continued as well. California policy makers implemented feed-in tariffs, providing financial incentives for new renewable resources, including solar and geothermal. Bonneville Power Administration offered long-term power purchase agreements to wind project developers, creating a stable wind-energy marketplace. Legislators in twenty-nine states, the District of Columbia, and three territories have set renewable portfolio standards.[6] For example, New Jersey requires that electric utilities supply 35 percent renewable energy by 2025 and 50 percent by 2030.[7] Fossil or nuclear as usual is no longer permissible.

These days, many conscientious homeowners and business owners willingly pay a little extra for green power as part of their electric bill.[8]

Wind power has matured at the utility scale as a cost-competitive technology (vis-à-vis coal and nuclear). The wind industry provides 120,000 jobs in the United States, according to the American Wind Energy Association.[9] Wind power now provides more than 20 percent of the electrical power in six states: Kansas, Iowa, Oklahoma, North Dakota, South Dakota, and Maine. The largest US wind farms are now in Texas, with more than 14,700 turbines.[10] There is a reason why the American Midwest has been called the Saudi Arabia of wind.

Both of the world's most populous countries take wind seriously. The sprawling 1,500-megawatt Muppandal Wind Farm in India blends with the human habitation in its midst.[11] The Jiuquan Wind Power Base (also known as the Gansu Wind Farm Project) in China's Gansu Province is, at 5,160 megawatts, by far the largest wind operation in the world. A transmission line is being built to take the power to China's central and southern cities.[12]

Advances in technology now make reliable offshore wind farms a reality. When the London Array was commissioned in 2013 at 630 megawatts capacity, it was one of the world's largest offshore wind farms. Now the United Kingdom has three offshore arrays that are larger. Northern Europe hosts forty-eight of the world's fifty-eight largest offshore wind farms. China hosts the other ten.[13] The WindFloat Atlantic floating wind project, connected by twelve miles of cable to the Portuguese coast, demonstrates the feasibility of developing wind power in waters up to one thousand feet (three hundred meters) in depth.[14]

Such developments helped Germany reach its goal, set in 2010, of meeting 35 percent of its gross energy consumption with renewables by 2020. In 2018, it achieved 38 percent![15]

SOLAR ELECTRICITY

Albert Einstein won his Nobel Prize for explaining the photovoltaic effect. Ever since, scientists and laypeople alike have dreamed of directly generating electricity from the sun. Unfortunately, it is hard to make an efficient solar electric system at a low cost. Fortunately, the attributes of no refueling and low maintenance compelled NASA to pay whatever it cost. During the 1970s, 1980s, and 1990s, remote installations provided a modest earth-based market for solar cells—freeway signs, mountaintop repeaters, cabins . . .

Along the way, researchers advanced solar cell technology. In 1977, when I studied solar energy in college, the efficiency of the best solar cell in normal sunlight was 14 percent. With concentrated sunlight, 22 percent efficiency could be achieved.

Fast-forward to 2020. A concentrator with layered cells to take advantage of the full solar spectrum yields a 47.18 percent efficiency. Just as significant, the best amorphous thin film solar collector, which can be made in large sheets, is 23.4 percent efficient.[16]

According to the Solar Energy Industries Association, in 2019, "The cost to install solar has dropped by more than 70% over the last decade. . . . An average-sized residential system has dropped from a pre-incentive price of $40,000 in 2010 to roughly $18,000 today, while recent utility-scale prices are competitive with all other forms of generation."[17] Meanwhile, the inflation-adjusted cost of new fossil-fueled generation has held steady.[18]

Declining cost and proactive policy complement each other. The Energy Policy Act of 2005 requires all electric utilities to accept power generated from rooftop collectors (or other generators) at retail rates.[19] The number of rooftop collectors installed on US homes doubled from one million in 2016 to two million in 2019.[20]

Utility-scale solar plants are also sprouting up in the desert. The Desert Sunlight Solar Farm, dedicated on February 9, 2015, east of Palm Springs, California, produces 550 megawatts of power, enough to power 180,000 homes. Located on federal land, it uses eight million photovoltaic panels. The similar-scale Topaz Solar Farm, on private land east of San Luis Obispo, California, has wide wildlife corridors between arrays. These projects are fostered by industrial-scale manufacturing processes, a federal loan guarantee, utility power purchase contracts, and a California mandate to produce a third of its electricity from renewable resources by 2020 and 50 percent by 2030. That said, large solar electric arrays have been built in Arizona, Nevada, and Texas as well.[21]

Combining residential- and utility-scale solar, 2.6 percent of total US energy production came directly from the sun in 2019.[22] Honduras leads the world in meeting 14.8 percent of its energy needs with photovoltaics. Israel, Germany, and Chile all come in at over 8 percent. In terms of sheer quantity, China is installing the most solar systems by far. Solar electric capacity was growing exponentially until 2017, when sales leveled off. That said, most systems stay in place for the long run. Three percent of the world's energy is provided by photovoltaics. Without those systems, our CO_2 emissions would have been 720 million metric tons higher.[23]

According to the United Nations International Energy Agency Photovoltaic Power Systems program, "32 countries had at least 1 GW [gigawatt] of cumulative PV capacity at the end of 2018 and 10 countries installed at least 1 GW in 2018." ("Capacity" denotes peak output. A gigawatt can power 110 million LED light bulbs.) In 2019, the world added over 120 GW *in one year.*[24]

Reporters for *Deloitte Insights* assert that solar and wind sources are moving from "mainstream" to "preferred" new energy technologies for investors.[25] Worldwide, renewable energy (including hydropower) is growing at around 8 percent per year. The International Renewable Energy Agency reports, "Nearly two-thirds of all new power generation capacity added in 2018 was from renewables, led by emerging and developing economies."[26]

Caution. Photovoltaic and wind systems *do* have an impact on their local environments. These include disruption of desert ecosystems and bird kills. Most sources of renewable energy have some impact. That is why conservation is described first in this chapter of earth-friendly energy resources.

DISTRIBUTED ENERGY

In 2002, Amory Lovins published *Small Is Profitable: The Hidden Economic Benefits of Making Electrical Resources the Right Size.*[27] It contained 207 ways that utilities and energy planners can better use our experience with renewable energy and energy conservation. In utility vernacular, these mostly environmentally benign strategies are referred to as distributed energy resources (DER).

Austin (Texas) Electric, among other utilities, is putting the principles described by Lovins into practice.[28] A purchase power agreement with a solar farm in West Texas offers economies of scale. Homeowners and municipal and commercial customers receive incentives to put solar collectors on their rooftops.

Austin actively partners with customers during times of peak demand on the electric system. Residential customers receive special space-conditioning thermostats and water heater timers. For commercial and institutional customers, Austin offers a load cooperative program, paying them to curtail part of their power demands in two-hour increments with one hour's notice. Another program coordinates the availability of backup generators used by hospitals and other regional entities. Large numbers of generators can be turned on with ten minutes' notice. While these generators use carbon-emitting diesel fuel, they are only on for short periods, and they reduce the need to build bigger power plants.

As part of its demonstration and development activities, Austin is experimenting with large lithium batteries to briefly provide power during short surges in demand and is demonstrating "cool storage" in which ice or stored cold water can help meet a commercial building's air-conditioning needs during the hottest hours of the day.

A December 2017 *Utility Week* article reports that Europe is also embracing distributed energy resources. Europe is now demonstrating virtual power plants, which automatically aggregate different kinds of distributed energy resources depending on need and availability. Research analyst Roberto Rodriguez Labastida asserts, "While the transition to DER will not be easy for organizations in an industry built around a centralized energy model, it is already possible to see the first sprouts of DER investments: a cleaner, cheaper, customer-focused, and far more innovative power sector."[29]

Utilities and regulators are beginning to see the growing fleet of electric vehicles as a resource. For instance, incentives can be offered to charge your car when solar electricity is available while you work instead of burdening the grid during peak evening hours.[30]

The solar cooker is also a distributed energy resource. This modest technology responds to all seventeen sustainable development goals (chapter 7 and appendix 2). Women and children's health improve when not exposed to smoke from a cooking fire (SDG 3 and 5). Children, freed from gathering wood, can go to school (SDG 4). Deforestation is abated and less carbon dioxide is released into the atmosphere (SDG 15 and 13). According to Solar Cookers International, more than four million solar cookers are in use, and three *billion* people rely on open solid fuel fires for their meals. The Handprint potential is huge!

Geothermal, energy from the earth itself, releases no carbon into the atmosphere. Ground-source heat pumps are used in homes and commercial buildings. Buildings in Boise, Idaho, and Klamath Falls, Oregon, are heated with geothermal water. Large geothermal power plants can be found in Iceland, Indonesia, the Philippines, and Northern California (the Geysers). There are issues: Geothermal electricity generation, with current technology, is limited to specific places on the planet. Geothermal power plants may see declining energy output over time. Technology used to increase potential may affect the groundwater.

And there are other plausibly eco-friendly energy technologies on the horizon. NovaSolix is using nanotechnology to go beyond the theoretical efficiency limits of semiconductor solar collectors.[31] Tidal power generates electricity at La Rance Station in France. PacWave, a grid-connected wave energy test and demonstration center near Newport, Oregon, is being developed by Oregon State University and the National Renewable Energy Laboratory.[32] Harnessing fusion power, as does the sun, is the goal of at least twenty-two US research sites and thirty-five collaborating nations using the ITER Tokamak (formerly the International Thermal Nuclear Experimental Reactor) in southern France.[33] One way or another, there are energy Handprints yet to be made.

HANDPRINT OPPORTUNITIES

Earth-Friendly Energy

14.1 Buy renewable power from your utility company.

14.2 Support your state citizens' utility board to advocate for eco-friendly power choices. Illinois, Wisconsin, Minnesota, and Oregon have such boards. Start one if you do not.

14.3 Work for a wind, solar, or efficiency company. Help others create Handprints.

Efficiency

14.4 Promote energy efficiency in your home and office. (See chapter 12 for ideas.) A smaller load makes it easier and less expensive to satisfy the remainder with renewable energy.

14.5 Donate to market transformation organizations such as the American Council for an Energy Efficient Economy, the Institute for Market Transformation, the Rocky Mountain Institute, or Natural Capital Solutions.

Wind

14.6 Share the land you love with windmills. Celebrate the intrinsic beauty in renewable energy.

Solar

14.7 Buy and use a solar cooker. Start with something simple, like making applesauce. Make solar cool by serving friends and neighbors' kids a solar lunch. Help Solar Cookers International bring clean energy to families that cook with wood and charcoal.

14.8 Install solar collectors on your home, either solar electric or solar hot water or both. Use net metering to feed your surplus electricity back into the grid and displace fossil fuels and nuclear power.

14.9 Go "community solar." Join with neighbors to pool financial resources and use available roofs and land for solar systems. If your state does not sanction community solar, as Oregon does, advocate for it.

14.10 Use solar-powered devices such as yard lighting and road sign lighting.

IN YOUR JOURNAL

What steps will *you* take to accelerate our adoption of earth-friendly energy?

VI

STEWARDSHIP HANDPRINTS

15

PROTECT AND
RECOVER THE LAND

In the late 1400s, the island of Hawaii developed a system of land owner-ship that respectfully took the natural world into account. *Ahupua'a* were narrow slices of land, typically bordered by ridges and extending from the highest ridge or mountaintop to the sea. The diversity of zones contained within the land allowed families to farm the midland areas, harvest fish from the sea, and harvest trees for canoes from the uplands. The highest forest, the *wao wakua*, where it was said the spirits lived, was relatively free from human intervention. The *ahupua'a* was expected to be self-sufficient—in other words, sustainable.[1] Early Hawaiian sustainable behavior was further motivated by *kapu* (taboos) that regulated fishing and harvesting (and social interaction). *Kapu* were enforced by *konohiki* and *kahuna* (priests).[2]

We know how to take care of the land when we choose to do so. Protecting the land before we "develop" it, and recovering it after we have "used" it, are both Handprints. Calling visionaries, politicians, regulators, biologists, trail builders, and ivy pullers to join these efforts.

THE SERENGETI OF AMERICA

Here is a story about what it can take to protect the land.

The area of northeastern Alaska, between the Brooks Range and the Arctic Ocean, is a place that defines "tundra," being home to black, griz-zly, and polar bears; caribou; foxes; wolves; the Pacific loon; and the buff-breasted sandpiper. Supreme Court justice William O. Douglas called the region the Serengeti of America, alluding to the abundance of wildlife in Africa.[3] Before oil was discovered in the 1950s, conservationists A. Starker Leopold, Olaus and Mardy Murie, Howard Zahniser, and David Brower wrote articles in national magazines and initiated letter-writing campaigns to preserve the great northern wilderness. In contrast, the first governor of Alaska advocated that the area be released to state control for drilling and mining.

The balance was likely tipped by Sigurd Olson, who visited northern Alaska in the summer of 1960 and then lobbied in Washington, DC, for its preservation. On December 6, 1960, interior secretary Fred Seaton designated 8.9 million acres (3.6 million hectares) as the Arctic National Wildlife Range, probably—we do not know for sure—with the blessing of President Dwight Eisenhower.[4]

Sigurd Olson went on to coauthor the Wilderness Act of 1964. While other threads also led to that momentous act, protecting of the tundra provided proof of concept.

The Handprint saga continues. The Clinton administration more than doubled the size of the refuge to 19 million acres (7.7 million hectares) and gave it a new name, the Arctic National Wildlife Refuge.[5] In 1991, a letter-writing campaign stopped drilling in the refuge. This despite the fact that the Senate Energy and Natural Resources Committee had already voted, by a margin of seventeen to three, for an energy bill that would allow drilling.[6]

On July 29, 2015, thirteen Greenpeace protesters rappelled down ropes from a bridge in Portland, Oregon. Their immediate objective: to block the passage of the icebreaker MSV *Fennica*, which was heading to the Arctic Ocean to help Shell drill for oil.[7] The whole country witnessed this aerial version of street theater. Two months later, at 1:00 a.m. on September 28, 2015, Royal Dutch Shell announced that it was abandoning plans to drill for oil in the Arctic Ocean. The company cited the poor performance of a test well. Analysts cited temporarily low oil prices. Politicians cited an unpredictable regulatory environment.[8] Who knows the inside story of this $3 billion write-off in favor of the environment?

ENDANGERED SPECIES

Protecting endangered wild animals generally protects the land on which they walk (or the water in which they swim, chapters 16 and 17). The United States is part of a broad international consensus that we should protect the endangered. Arguably the grand master of environmental agreements is the Convention on International Trade of Endangered Species of Fauna and Flora (CITES, pronounced site-ease). This is the world's response to the international trade of elephant tusks and wild animal pelts that was driving many animals to near extinction. The term "tickles the ivory" refers to the fact that piano keys were made from elephant tusks. The United Nation resolution recommending such a treaty was adopted in 1963. The treaty opened for signatures in 1973. It entered into force on July 1, 1975.[9] Twelve years of continuous negotiation. Actually, it took more like four decades if you refer back to the London Convention Relative to the Preservation of Fauna and Flora in their Natural State, which listed forty-two species in 1933.[10]

CITES has three appendixes of protected wildlife, which are periodically updated. Commercial trade is prohibited for 1,200 species in its appendix 1. This is where you will find the iconic species that are threatened with extinction—African and Asian elephants, the African lion and other big cats, the gorilla and the chimpanzee, most rhinoceroses, and the manatee. The treaty's appendix 2 lists 21,000 species that are subject to strict regulation, including the American black bear, the African gray parrot, and the big-leaf mahogany tree. About 170 species are listed in CITES's appendix 3 at the request of one country that requests help of other countries in protecting a given species. The United States asked for help in protecting the alligator snapping turtle in southeastern states. This "dinosaur of the turtle world" can grow to two hundred pounds.[11] Costa Rica did the same for the two-toed sloth.[12]

CITES intervenes in the market. It does not prosecute individuals or countries. That is up to individual nations. The measures that are available include confirmation of permits, suspension of cooperation, recommendation to participating nations to suspend trade, and specification of measures that would be needed to resume cooperation. Now most of the members of the United Nations (including the United States, the Russian Federation, and China) are signatories of CITES.

A nation meets CITES "accession" requirements when it passes corresponding internal legislation. In 1992, the Czech Republic needed to pass its own legislation authorizing CITES in anticipation of joining the European Union.[13] China fulfilled its 1981 accession to CITES with the Law for Protection of Wild Animals in 1989 and the Law for Protection of Wild Plants in 1997.[14] In the United States, the Lacey Act, passed way back in 1900, contains a unique provision that makes it unlawful to trade wildlife or wildlife products acquired illegally in another country.[15]

US participation in the international CITES processes is guided by the Endangered Species Act of 1973, with its own criteria for listing a species.[16] This law and those who enforce it have allowed more than a hundred species to recover from near extinction, including the Aleutian Canada goose, the black-footed ferret, the American crocodile, the whooping crane, the gray wolf, and the shortnose sturgeon. Adding the spotted owl to the act's endangered species list gave biologists, tree huggers, and ultimately the courts an indicator species for the health of northwest forest ecosystems.[17] This allowed us, starting around 1990, to curtail the harvest of the remaining temperate ancient forest in the Pacific Northwest.

PERMACULTURE

Conversely, there is still room for Handprints on land occupied by humans.

Bill Mollison and David Holmgren, of Australia, proposed a set of principles to create a healthy environment. Their coauthored book, *Permaculture One*, published in 1978, recommends that each design feature have at least three functions (a lawn only serves an esthetic purpose). Holmgren's *Permaculture: Principles and Pathways Beyond Sustainability*, written in 2002 with a quarter century's additional experience, describes how permaculture thinking applies to whole landscapes—and grounded strategies to change systems.

SHAPING A HEALTHY SPACE, OBSERVATIONS BY DAVID HOLMGREN

- Learn to read the land to reduce the need for fossil fuel and labor. "An open inquiring attitude to problems is almost always more fruitful than an urgent demand for solutions" (pp. 13, 18).
- Energy from the sun is stored in a landscape in the form of water, living soil, trees, and seeds. Food is the most important form of energy. Culture is also a form of energy storage (pp. 27, 46).
- Any project that does not yield benefits—nutritional, financial, or esthetic—will not be well maintained. Selecting hardy, locally adapted and self-reproducing plants can minimize the resources required to maintain gardens, farms, and forests (pp. 55, 59).
- "If our design and care are truly inspired, the garden, as well as becoming more robust, will develop a degree of self-

regulation and balance analogous to children growing through adolescence to adulthood" (p. 74).

Source: David Holmgren, *Permaculture: Principles and Pathways Beyond Sustainability* (Hepburn, Australia: Holmgren Design Services, 2002). The first four of twelve principles are highlighted here.

One community in Sebastopol, California, typifies the practical application of permaculture principles.[18] A group calling themselves Permaculture Artisans converted a downtown lot of pavement and gravel into a showcase permaculture garden. It features two pools, eighty fruit trees, edible mushroom cultivation, 40,000 gallons (150,000 liters) of rainwater harvesting and storage, graywater integration from an outdoor kitchen, and a flagstone patio. Four adjacent neighbors removed their fences and assist in the upkeep of the garden. The artisans give private tours of the garden. In this case the "yield" is income from clients who also wish to practice permaculture.

The Permaculture Worldwide Network website reports over 2,600 permaculture projects around the world, including two in Mongolia.[19] The Permaculture Research Institute offers a Permaculture Teacher Registry, complete with application process, to ensure that the principles of permaculture are widely available while maintaining credibility.[20]

Food Not Lawns, founded in Eugene, Oregon, in 1999, employs permaculture principles to weave sterile surroundings into healthy landscapes.[21] Heather Flores's experience in turning parking strips and side yards into gardens yielded the 2006 book *Food Not Lawns: How to Turn Your Yard into a Garden and Your Neighborhood into a Community*. Heather Flores's 2015 Edible Nation Tour provided down-in-the-dirt help organizing local groups. Over fifty chapters across the United States, Canada, and Australia have formed.[22] That is a Handprint!

The Portland Audubon Society offers Backyard Habitat Certification criteria. This inspires and guides homeowners to eliminate invasive plants, establish up to five canopy levels, and implement storm water management.[23] The National Wildlife Federation offers a similar program throughout the United States.[24]

BIOREMEDIATION

Handprints can also be created out of commercially and industrially degraded land. Oil spills, old gas stations, and chemical dumps really can be brought back to life. Bioremediation helps us decontaminate brownfield sites using biologically friendly methods. Techniques include injecting oxygenated water into the groundwater; introducing microorganisms that biologically break down pollutants; and stimulating natural microorganisms with nutrients, temperature, or moisture.[25]

Another practical approach to bioremediation is described in the pioneering book *Mycelium Running*, published in 2005 by Paul Stamets. Mushrooms (mycelium) can turn a hydrocarbon-soaked landscape into ecologically friendly carbohydrates. In an oft-reported study hosted by Battelle National Laboratory, four piles of diesel-soaked earth were observed over time. One pile was left untreated as a control, one was treated with enzymes, one was treated with bacteria, and one was inoculated with oyster mushrooms. Within six weeks, when tarps were removed, the mycelium-inoculated pile was covered with mushrooms. Within eight weeks, green plants had recolonized the pile that grew mushrooms. The other piles remained dark and lifeless. The mushrooms dropped the aromatic hydrocarbon count from 100,000 to 200,000 parts per million. Stamets and his associates now place burlap sacks of mycelium-inoculated wood chips near farms and factories to neutralize harmful bacteria and reduce coliforms.[26]

Stamets's work has spawned a bioremediation specialty: mycoremediation.[27] Mushroom Mountain, out of South Carolina, is helping clean up toxic waste pools left by decades of oil drilling in the Amazon.[28] CommonWealth Urban Farms in Oklahoma City are using elm oyster mushrooms (*Hypsizygus ulmarius*) to recover a contaminated city lot. From their website: "Research has shown that mushrooms combine lead with phosphorus creating chloropyromorphite, the most stable lead mineral. Mushrooms also break down pesticides like Chlordane into their base components."[29] Researchers from the World Agroforestry Centre and the Kunming Institute of Botany in China discovered a fungus that breaks down plastics in landfills.[30]

In 2016, Peter McCoy wrote *Radical Mycology: A Treatise on Seeing and Working with Fungi*, which includes mycoremediation. He now offers mycological education online via Mycologos. The vision is to create and support "a *mycoculture*: a global community of highly educated fungal advocates, cultivators, and allies working to apply this knowledge for the greatest good."[31]

Over 430 bioremediation companies, listed on the Environmental Expert website, offer an amazing diversity of services.[32] JRW Bioremediation, based in Kansas, uses environmentally friendly products to treat ground-water contamination from chlorinated solvents, metals, and perchlorates. Scipio Biofuels, based in California, uses proprietary algae processes to treat water bodies that are so rich in nutrients that aquatic animals cannot get oxygen. RNAS Remediation Products, based in Minnesota, offered the first off-the-shelf emulsified vegetable oil for bioremediation. Their "crop-derived electron donors" treat soil and groundwater while minimizing their carbon footprint. Waterstone Environmental, based in Colorado, is a woman-owned business that provides mine waste and heap leach remediation, aerial photographic surveys, and LANDSAT image interpretations.

This bioremediation capability does not happen by accident. Members of the Waterstone team graduated from five different universities and obtained extra certificates beyond the coursework.[33] Studies by the US Geological Survey amass the knowledge taught in the universities.[34] Funding for those studies comes from a tax on chemical and petroleum companies that was authorized by the Comprehensive Environmental Response, Compensation and Liability Act of 1980 (CERCLA)—known as the Superfund.[35]

THE SUPERFUND

With the passage of CERCLA, the federal government gained the authority to assess liability and direct site cleanup of abandoned and uncontrolled hazardous waste dumps.[36] When a site is listed on the National Priorities List, it becomes subject to Environmental Protection Agency oversight and eligible for cleanup from the Superfund. Potentially responsible parties provide financial help and cooperation as they are motivated and able to do so. The National Oil and Hazardous Substances Pollution Contingency Plan guides the overall response. The oft-updated plan was originally developed to respond to the 1968 *Torrey Canyon* tanker oil spill in 1968.[37]

Many formerly toxic Superfund sites—especially large and contaminated brownfields—have been redeveloped to serve human needs in a more environmentally benign way.[38] Potentially responsible parties around Leadville, Colorado, cooperated to clean up eighteen square miles of mining and smelting wastes. A more diversified economy is now being fostered by tourism. The Quapaw Tribe of Oklahoma removed 108,000 tons of contaminated mining waste to provide archeological, educational, and historic preservation opportunities near a closed boarding school. Worcester Township, Pennsylvania, hosts a clean international distributer of hand tools and soldering equipment after detoxifying decades of industrial operation. One hundred and thirty-five acres of the York County, Pennsylvania, solid waste landfill was transformed into the Hopewell Area Recreational Complex with trails, wildlife viewing platforms, athletic fields, a picnic platform, and a two-acre solar array to power ongoing groundwater treatment. The former Brick Township landfill, in New Jersey, is now graced with 24,000 solar panels.

Some Superfund sites actually feed us.[39] Farming and ranching simply continues—in a healthier way—at Superfund sites from Vermont to Nevada. Philadelphia's defunct Boyle Galvanizing Plant is now Greensgrow Farm, host of a community-supported agriculture program and recycler of vegetable oil from local restaurants.

Citizen action is often integral to rehabilitating a Superfund site. Andrea Amico led efforts to galvanize concern about water pollution from Peace Airforce Base in New Hampshire.[40] The Portland (Oregon) Harbor Coalition, which includes urban Native Americans and immigrant families, drew in 5,300 comments regarding a plan to clean up eleven miles of the Willamette River.[41]

At least seventy Superfund sites have been returned to the environment: The Alabama Army Ammunition Plant is now a wildlife preserve. The Great Swamp National Wildlife Refuge was, in part, a New Jersey asbestos dump. The Chemical Commodities site in Kansas provides habitat for local wildlife and pollinators. Land polluted by E. I. Dupont De Nemours & Co. in Delaware was restored as a twenty-acre wetland and pollinator meadow. The polluted drinking water well field of Elkhart County, Indiana, became a botanical garden.

Since the passage of the Superfund Act, over 1,700 toxic sites have been placed on the National Priorities List. Of these, remediation ("construction") has been completed for over 1,200 sites. Often the site returns to usefulness, even while long-term cleanup actions continue. This is expensive, complicated, and large-scale progress.[42]

This is how we clean up a nation, a particularly necessary form of collective Handprint.

HANDPRINT OPPORTUNITIES

Land Protection

15.1 Visit a national park. Bring children. Play by the rules. Help the National Park Service succeed in its dual mission of protecting nature and making it accessible. Find out how the park was created and its needs. Inspire others.

15.2 Support trail building and maintenance. Make it easy to love nature.

15.3 Support indigenous people who protect the land, including Honor the Earth (Great Lakes region of North America), the Pachamama Foundation (the Amazon and the Andes), and the Rainforest Action Network (worldwide).

15.4 Donate to African Parks to help it manage—with national governments and local communities—nineteen national parks in Angola, Benin, Central African Republic, Chad, the Democratic Republic of Congo, Malawi, Mozambique, the Republic of Congo, Rwanda, Zambia, and Zimbabwe.[43]

15.5 Bequeath your land to a conservation organization like Western Rivers or Nature Conservancy.

15.6 Form a land trust where nature can flourish. Insist that developers set aside a commons that will serve both neighbors and nature.

15.7 Support organizations that protect endangered species. Check out "15 Actions to Protect Endangered Species" on the Endangered Species Coalition (US) website.[44] The Nature Conservancy and the World Wildlife Federation, among others, stand for endangered species worldwide.

Land Recovery

15.8 Create a certified wildlife habitat out of your balcony, yard, roadside, schoolyard, or workspace. Use the criteria of the National Wildlife Federation.[45]

15.9 Use permaculture principles to evolve beyond lawns and parking lots. Plant fruit trees, berry bushes, and vegetables. Build benches.

15.10 Use mycelium (mushrooms) to treat biohazards of farms, factories, and city lots. Check out *Mycelium Running* by Paul Stamets for ideas.

15.11 Help expand the fungi knowledge base. Check out the North American Mycoflora Project, a citizen science opportunity. Contribute observations to iNaturalist, which is coordinated by the California Academy of Sciences and National Geographic.[46]

15.12 Promote brownfield recovery—as a public citizen, a cooperating potentially responsible party, or a developer. For examples, refer to "Anatomy of Brownfields Redevelopment," a 2006 pamphlet published by the Environmental Protection Agency.[47]

15.13 Adopt a park with love and long-term commitment. Use your back, your professional skills, and your money as appropriate. Check out Grand Staircase Escalante Partners (Utah), Yosemite Conservancy (California), and Central Park Conservancy (New York City).

IN YOUR JOURNAL

What steps will *you* take to protect and recover the land?

16

REVIVE OUR RIVERS

One day every year, the Johnson Creek Watershed Council hosts a dozen simultaneous activities in a watershed-wide event along the creek's twenty-six-mile (forty-two-kilometer) reach.[1] Portland, Oregon, and its neighboring communities have been cleaning up Johnson Creek for a quarter century. Besides removing sodden blankets, car tires, and shopping carts, volunteers plant trees to lower the temperature of the water with fish-friendly shading. After decades of absence, a few salmon have returned.

When rivers are unhealthy, the land is unhealthy, and people become unhealthy too. Reviving a river that has been polluted for decades is some of the toughest work we can imagine. It requires a cast of thousands, perhaps millions. Such collective Handprints have been created again and again.

PETE SEEGER'S LEGACY

The Hudson River runs south from upstate New York to meet the Atlantic Ocean at New York City. Its banks reflect the growth of industry in the nation. There was a time when the General Motors plant at Tarrytown painted the river the color of its cars, and the General Electric plants in Washington County dumped polychlorinated biphenyls (PCBs) from making transformers into the river. Meanwhile, Manhattan released 150 million gallons of raw sewage into the river. By the mid-1960s, much of the river could no longer support fish. The river had died.

Pete Seeger, a well-known singer who lived on the banks of the Hudson, was well aware of these challenges. He joined with others to build a

106-foot (32-meter) sailboat, the kind that plied the river in the more bucolic days of the eighteenth and nineteenth centuries. The intention: draw people to the river and help them care about it. From its 1969 launch, the sloop *Clearwater* partnered with schools and environmental organizations along the river.[2] The annual Clearwater Folk Festival started as a fundraising effort to build the boat. Pete Seeger's singing was part of the allure, and so was the opportunity to visit the past and learn how to sail. More than half a million people have seen the river from a new perspective aboard the *Clearwater*. Over time, the sloop has become a training platform for activists.

Pete Seeger's concerts and the *Clearwater* fueled a growing concern for the environment. In 1965, New Yorkers passed a billion-dollar bond to fund sewage treatment. Industry gradually cleaned up its act—with the help of the 1972 Clean Water Act and the awareness of the public.[3] More recently, circa 2012, forty-two New York counties and stewardship organizations helped create and endorsed the Mid-Hudson Regional Sustainability Plan.[4]

The Hudson job is not done. A century's accumulation of toxic sludge (read: PCBs) still needs to be dredged.[5] Pregnant women and children are still discouraged from eating fish from the river. But, fish and humans once again swim in the Hudson River.

The Waterkeeper Alliance, started in 1966 by blue-collar fishermen, also helped motivate the Hudson River cleanup. This effort has catalyzed some 340 Waterkeeper organizations and affiliates in forty-four countries. According to their 2018 annual report, "Waterkeeper Alliance ensures that the world's Waterkeeper groups are as connected to each other as they are to their local waters, organizing the fight for clean water into a coordinated global movement."[6]

REMOVING DAMS

On October 28, 2011, the siren before the breaching of the Condit dam on Washington's White Salmon River was reminiscent of an air raid alarm. Deep within the dam, a series of explosions rumbled. Then black water gushed from the dam's base. Like a giant fire hose, the roiling water scoured everything in its path downstream.[7]

Such violence is strong medicine to bring a dammed river back to its natural state, to once again allow fish to migrate. It is also testimony that we really are capable of dismantling edifices that no longer serve the greater good. For this to happen, indigenous people have led decades-long efforts to align politics, economics, permit expirations, science, and treaty rights.

Dams have been legally removed on Elwha River in Washington state, Siletz River in Oregon, Mill River in Massachusetts, and Uwharrie River in North Carolina.[8] Since 1912, over 1,600 dams have been removed in the United States, including 99 in 2018, up from 91 in 2017.[9] The Edwards Dam was removed from Maine's Kennebec River in 1999. It was the first functioning hydroelectric dam to be removed in the United States; the economic value of a restored fishery was shown to be greater. Along with the sturgeon, alewives (a type of herring), and shad came river otters, bears, minks, bald eagles, ospreys, and blue herons.[10]

It is a work in progress: removal of four dams on the Klamath River in southern Oregon awaits a $250 million congressional authorization for river restoration and economic development. Mission impossible? At least the long-bickering ranchers, farmers, Klamath Indians, and Pacific Power are on board.[11]

WATER RESTORATION

Each American's embedded water Footprint is two thousand gallons per person per day.[12] That accounts for our food, our appliances, and our military. Just so you know.

We can actually help put water back into a dry riverbed. By purchasing a Water Restoration Certificate for two dollars from Bonneville Environmental Foundation (BEF) you restore a thousand gallons of flow to a stream somewhere in the American West.[13] How does this work? BEF partners with organizations like Change the Course, the Nature Conservancy, and Intel to field on-the-ground projects with local participants. Strategies include water conservation from new irrigation systems, water management including using existing irrigation systems more efficiently by reducing leaks and monitoring soil water content, water leasing to leave water in streams that could legally be removed, and permanent transfer of water rights. As with carbon offsets, independent companies verify that the work actually gets done. And look what we end up with.

The discharge from Mexicali's Las Arenitas wastewater treatment plant was intended to serve agriculture. But, to put it nicely, the discharge water did not meet health standards. Funding from Coca-Cola and White Wave Services (makers of Silk soymilk) and help from the Walton Family

Foundation allowed the construction of a 250-acre (100-hectare) wetland, which provides additional water purification. Twenty million gallons (76,000 cubic meters) of high-quality water per day now flow into the Rio Hardy, which flows into the Colorado River Delta. The result: a sanctuary has been restored for the yellow-rumped warbler, American coot, and other birds.[14]

Restoring natural flow from a diversion on Jesse Creek in Idaho will re-create a forty-acre (sixteen-hectare) wetland and add cutthroat trout habitat in the greater Yellowstone ecosystem.[15] The city of Aspen, Colorado, implemented water conservation sufficiently to free up one of its "senior water use rights" to the Roaring Fork River. Patagonia, Clif Bar, New Belgium, and Ted's Montana Grill pitched in to help fund the project.[16]

New drip irrigation systems annually restore up to seven million gallons (twenty-seven thousand cubic meters) to Arizona's Verde River. The Nature Conservancy is a leading partner on this project.

Intel, with an ongoing need for lots of clean water for microchip production, takes a special interest in water restoration and is involved in more than a dozen projects. With Intel funding, New Mexico's Comanche Creek has been reconnected with its natural floodplain to restore natural water storage, part of a larger initiative to restore 109 million gallons of water per year to the Rio Grande River.[17] Eleven Intel-funded projects now restore more than a billion gallons of water every year to naturally flowing rivers.

Water restoration over seven states now totals more than 13.5 billion gallons (51 million cubic meters).[18] The objective of water restoration projects is not to deliver water to the sea; often the lower stretches of desert rivers remain dry. The real value of restoring water to streams and rivers is healthier habitat—measured in acres and miles (hectares and kilometers) of river and protected species.[19]

ICONIC RIVER REVIVALS

The Cuyahoga River, which flows into Lake Erie at Cleveland, Ohio, caught fire from floating oil and debris at least half a dozen times. The biggest fire was in 1952, but the last fire in 1969 proved to be yet another rallying cry for the Clean Water Act of 1972.[20] The establishment of Cuyahoga Valley National Park, in 2000, provides de facto water filtering and spawning grounds between the industrialized cities of Akron and Cleveland. The $200 million Mill Creek tunnel, created by the Northeast Ohio Regional Sewer District, ensures that sewage does not contaminate the water during storms.[21] Today, more than forty species of fish swim in the river that used to catch fire.[22]

As with the Hudson, there is a qualifier. Ohio advises that people not eat more than one meal of locally caught sport fish per week, statewide.[23] Mercury still accumulates in larger fish.

In 1986, Lewis MacAdams and two other people founded the Friends of the Los Angeles River (FoLAR) by cutting a chain-link fence separating the waterway from the people. At the time, it was a seemingly permanent concrete ditch. In 1989, ten thousand people participated in a "river" cleanup, which encouraged the public to visualize an actual river. At one point, FoLAR raised $1 million so the Army Corp of Engineers could complete a study to restore eleven miles of the river. By 2016, this resulted in the Los Angeles River Restoration Project to maintain flood risk management, enhance recreation opportunities, and reestablish river habitat that connects with the Santa Monica and San Gabriel Mountains. The city of Los Angeles is buying derelict lots along the "river" to eventually create a fifty-one-mile (eighty-two-kilometer) greenway. More work in the river is needed, but the people and relevant institutions are on board.[24]

The Mizaan Project, led by Nelley Youssef and fellow Muslim volunteers in Sydney, Australia, cleaned up a stretch of Cooks River, one of Australia's most polluted rivers. Once-barren banks again host native plants, blue-tongued skinks, and birds.[25] In the process of planting 4,500 trees over three years, the project helped create positive rapport between Australia's Muslim minority and the secular culture.

London's Thames River, once a dead river, is now home to 150 species of fish and a thriving otter population. The ninety-plus projects in recent years include "catchment sensitive farming," wetland creation, and reducing storm sewer overflows.[26]

A full-scale treaty supports the cleanup of the Danube River, one of the world's most international rivers. The Danube originates in Germany, passes through Vienna and Belgrade, receives water through tributaries from Bucharest and Sofia, and ends in a delta on the Black Sea. River pollution sources include agriculture, raw sewage, and mine spills. In 1994, eleven riparian (riverside) nations signed the Convention on Cooperation for the Protection and Sustainable Use of the River Danube (Danube River Protection Convention).[27] Danube Day is celebrated annually in many of the nineteen countries served by the river. The Green Danube Partnership provides the "Danube Box" to educate hundreds of thousands of children about water conservation and management with teaching aids, games, and posters. Financial support is provided by Coca-Cola.[28]

People from fifty-eight cities volunteer for the annual Rhine River cleanup day. A 1986 fire in Switzerland's Sandoz chemical plant turned the already heavily polluted Rhine River red in every country it passed through. That motivated France, Germany, the Netherlands, and Switzerland to agree to the Rhine Action Program. Its seemingly impossible goal: return the salmon to the Rhine by 2000. The resulting cooperation led to both strict industrial regulations and extensive investment in sewage treatment systems. Salmon returned to the river in 1997, three years ahead of schedule.[29]

HANDPRINT OPPORTUNITIES

16.1 Walk and play along streams with children. Tell them how the Thames, Danube, Hudson, and Cuyahoga Rivers recovered. The sheer scale of these cleanup efforts—and the profound results— give us hope. Play, stories, love, and stewardship are related.

16.2 Volunteer to help restore your local stream or watershed. Remove invasive species. Remove culverts that stop fish passage. Initiate citizen-based enforcement of environmental laws where regulatory authority is lacking. If there is no local restoration council, start one.

16.3 Disconnect your downspouts from the sewer to end sewer outfalls into the river.

16.4 Create rain gardens beside streets and parking areas to reduce runoff.

16.5 Join American Rivers in their campaign for fish runs and ecosystems by removing unnecessary dams.

16.6 Buy Water Restoration Certificates from BEF. Help save water ecosystems across the United States. Make your business water neutral.

16.7 Support the Ganges River cleanup. The Ganga Action Parivar's 6T's program is a recipe for a healthier India: *t*oilets for schools and communities, *t*rash (solid waste) management, *t*rees to mitigate erosion, access to clean water *t*aps, upgrading India's railroad *t*racks, and protecting *t*igers and other endangered wildlife.[30] Millions of hands, hearts, and dollars are needed to create this magnificent collective Handprint.

IN YOUR JOURNAL

What stream or river will *you* help revive?

17

CONSERVE OUR OCEANS

When the ocean is sick, the prognosis for our future is not good. However, the ocean has an amazing capacity to heal when we respect it, clean it up, and restrain our excessive habits. That which we do to heal the ocean, we do for everyone. We humans have an impressive track record of ocean Handprints on which to build. Moving outward from the shore . . .

THE COAST

In 1913, the Oregon legislature declared its Pacific coast a state highway. That act of public-spirited lawmaking resulted in a state park every ten miles. One day, decades later, a resort owner blocked off the dry part of the beach exclusively for its guests. Citizens Robert Bacon and Diane Bitte brought the issue to the attention of soon-to-be governor Tom McCall (a Republican). Reporter Matt Kramer picked up the story. KGW television manager Ancil Payne dramatized it. This set the stage for passage of the 1967 Beach Bill, which gave *everyone* free and uninterrupted access to 362 miles (582 kilometers) of ocean coastline below where permanent vegetation grows (legally, up to sixteen vertical feet above the low tide).[1] That bill was modeled after the Texas Open Beach Act of 1959, which guaranteed access to the Gulf Coast.[2] Hawaii ensures public access to its beaches as well.[3]

Internationally, the Trash Free Seas Alliance, led by the Ocean Conservancy, coordinates many beach cleanups. It has such diverse partners as Coca-Cola (lots of volunteers), Altria (disposal of cigarette butts), the Michigan State University School of Packaging, and the World Society for the Protection of Animals.[4] Janis Jones, CEO of Ocean Conservancy, says, "In 2017, nearly 800,000 volunteers collectively removed more than 20 million pieces of trash from beaches and waterways around the world."[5]

Discarded cigarette filters are ubiquitous. According to *National Geographic*, smokers buy some eighteen billion cigarettes *a day*, most of which have slow-to-degrade cellulose acetate filters. Two-thirds of all cigarette butts are flicked directly into the environment, and many of them make it to the beaches and the ocean. Millions of disposable e-cigarettes, each with a plastic tube and battery, further exacerbate the challenge.[6]

Back in Oregon, SOLVE coordinates beach cleanups in the spring and fall. My Portland congregation has fielded several crews. It's fun.

MARINE PROTECTED AREAS

In 1999, California passed the Marine Life Protection Act and subsequently set up a network of over 120 marine protected areas along its 1,100-mile (1,770-kilometer) coast. The goals were to conserve habitat; allow marine life to thrive; preserve natural diversity; help rebuild depleted populations; and offer educational, research, and economic opportunities.[7] Included are a few special closure sites adjacent to seabird rookeries and sea mammal haul-out sites.

The US federal government has set aside some three hundred Marine Protected Areas under various auspices. Of these, fourteen marine sanctuaries are operated by the National Oceanic and Atmospheric Administration, ranging from the Marianas Trench—the deepest in the world—to the Northwest Hawaiian Islands, and from Lake Huron to Gray's Reef off the state of Georgia's coast.[8]

Barack Obama added more than 850,000 square miles (2.2 million square kilometers) to our protected seascape, an area three times the size of Texas. This was done largely by extending existing Pacific sanctuaries, created by President George W. Bush, out to the 200-mile (320-kilometer) US territorial limit. Nearly a fourth of all the water we can protect is protected. No commercial fishing. This expansive Handprint gives endangered species hope of thriving once again.[9]

FISHERIES

By 1959, the fisheries in Alaskan coastal waters were collapsing—the catch was less than a third compared to the 1930s, and there were many more fishermen. Coastal towns were depressed. The need to control absentee corporate cannery owners fueled Alaska's drive for statehood in 1959. The resulting state constitution required "sustained yield" management of fish and wildlife. Shortly after achieving statehood, the Alaska legislature outlawed fish traps. In the following years, Alaska set in place management practices such as "escapement goals" for spawners to repopulate the fisheries, in-season closures, limited fleet size, and rules governing type of gear and location and times of fishing.[10] Clem Tillion, a fisherman, homesteader, and legislator, said, "Don't let me get between you and the fish; I'll vote for the fish every time."

Alaska senator Ted Stevens sponsored the Magnuson-Stevens Fishery Conservation and Management Act of 1976. Its purpose, according to the senator, was "to protect the reproductive capacity of our fisheries." Significant amendments in 1996 and 2006 were required to definitively accomplish that task.[11]

Internationally, fisheries management has gained traction as well. Some four hundred publications tell stories of fisheries from Micronesia to Ireland and from the Mediterranean to the South Indian Ocean. Steven Murawski of the National Marine Fisheries Service says, "Of the 24 depleted stocks worldwide with formal rebuilding plans to reduce excess fishing mortality, only one has not recovered."[12]

REEFS

Established in 1992, the Flower Garden Banks National Marine Sanctuary (FGBNMS) includes coral reefs in the northwestern part of the Gulf of Mexico, 119 miles (192 kilometers) southwest of Galveston, Texas. The area is *entirely submerged*, rising to within fifty-six feet (seventeen meters) of the water's surface. It hosts a robust colony of brain coral in the shallows and large treelike corals in the surrounding depths. The area is surrounded by oil and gas drilling rigs and connecting pipelines—*and* the reef is healthy.[13]

Did the April 20, 2010, blowout of the BP *Deepwater Horizon* oil platform and the subsequent Gulf oil spill damage this reef? Not this time. Surveys the following summer conducted by the FGBNMS and the larger Florida Keys Marine Sanctuary found no damage that could be directly linked to the Gulf oil spill. The *Deepwater Horizon* was 300 miles (483 kilometers) to the east of the FGBNMS. These surveys matter because the act creating marine sanctuaries and the Oil Pollution Act of 1990 both have teeth. Had a deep plume of oil, gas, and dispersants reached the sanctuary, BP could have been liable for a fine of *$200,000 a day* plus "the costs of actions authorized by the Secretary [of Interior] to minimize destruction or loss."[14] That could get a corporate executive's attention.

The state of the climate also endangers reefs. Warming water, driven by increasing carbon dioxide in the atmosphere, profoundly affects reefs. Coral lives in a symbiotic relationship with the algae called *zooxanthellae*, which are very small plants with photosynthesis capability. These algae feed the coral, which are actually small animals. When the water temperature rises, the algae can die, starving the coral. That is when the reef "bleaches."[15]

While landlubbers work to reduce the global flow of carbon into the atmosphere, the International Coral Reef Initiative, with sixty countries participating, coordinates efforts to reduce local stressors to reefs. These include creating marine sanctuaries and zoning reefed areas as no-fishing and no-anchor zones ("lines in the water"). Administrators even negotiate with oil companies to use abandoned drilling rigs as foundations for new reefs.

Another ally in our efforts to protect reefs is the sea otter.

SIGNATURE MARINE SPECIES

The playfulness of sea otters belies their special niche in marine ecology. Sea otters eat sea urchins, but apparently only enough to hold urchin populations at bay. This turns out to be hugely important because sea urchins, left unchecked, eat live kelp. Kelp is one of the key absorbers of carbon dioxide from the atmosphere, helping control acidification, coral bleaching, and ultimately global warming. In addition, kelp forests protect ocean beaches from erosion. Indirectly, sea otters also protect seagrass, which provides needed fish and turtle habitat.[16]

How are the otters doing? The southern sea otter population off California fell from about sixteen thousand to about fifty animals in the 1970s. The popularity of their pelts and oil spills were two of the contributing factors. By 2017, with "fully protected" status in place, the known population rose to 3,186 animals.[17] (They count that carefully.) Save the sea otter!

Moving from the cute to the magnificent . . . The International Convention for the Regulation of Whaling, signed in 1946, established the International Whaling Commission (IWC). The IWC now has representatives from eighty-nine member governments. The commission has long sought to "provide for the proper conservation of whale stocks" and make possible the continuation of whaling as an industry. The IWC operates much like a US state game commission, with designated seasons, criteria for what whales can and cannot be taken, designated sanctuaries, establishment of conservation plans for specific species, ongoing monitoring of the whale populations and the take, guidelines to avoid ship strikes, response plans for entangled whales, and ongoing diplomatic and scientific dialogue.[18] Despite the excesses of nonmember nations, the IWC has remained relevant for more than seventy years.

Is this effort to protect both the industry and the whales really succeeding? Preexploitation populations of whales worldwide numbered in the hundreds of thousands. A few species, like the humpback whale, in the mid latitudes, appear to have recovered to near preexploitation levels. Most whale populations are recovering from their unprotected levels but are nowhere near preexploitation levels. The gray whale has recovered in the Eastern North Pacific and has declined to only 130 animals in the Western North Pacific. The North Pacific and North Atlantic right whales, hunted mercilessly in the 1800s, are the only large whale species that are said to be extremely endangered.

One summary accomplishment: *no large whale has gone extinct under the watch of the IWC.*[19]

The Marine Mammal Protection Act of 1972 forbade nearly all taking of marine mammals in waters controlled by the United States. Some exceptions, particularly for indigenous people, were made in the act's 1994 amendment.[20] The Endangered Species Act of 1973, which included the protection of marine mammals, was even bolder. The Supreme Court found that Congress intended to "halt and reverse the trend toward species extinction, *whatever the cost*"[21] (emphasis added). Thus whales, porpoises, dolphins, turtles, manatees, sea lions, seals, and otters have been given opportunity to recover.

We, as a public, would not have known dolphins were being killed indiscriminately if Samuel LaBudde had not worked undercover for the Earth Island Institute. For three months, LaBudde played the role of an eccentric cook aboard the *Maria Luisa*, a tuna boat of Panamanian registry that shipped out of Ensenada, Mexico. He witnessed dolphins dying as bycatch, caught in purse seine nets. His videos, and the Earth Island campaign that followed, led to the Dolphin Protection Consumer Information Act of 1990 and the International Dolphin Program Conservation Act of 1997. To earn "dolphin friendly" labels on canned tuna, Bumble Bee, its competitors, and their suppliers cannot use purse seine netting. Another undercover filming project by the Earth Island Institute led to banning of even more indiscriminate and larger-scale open-ocean drift netting by a United Nations general resolution in 1992.[22]

Dr. Archie Carr first brought public attention to the decline of sea turtle populations with the publication in 1959 of *The Windward Road*. In particular, he called attention to the green sea turtles of Tortuguero Beach in Costa Rica. The Caribbean Conservation Corporation, which became the Sea Turtle Conservancy, was founded with the help of philanthropy.[23] The organization tried to "head start" turtles in hatcheries and distributed eggs around the Caribbean with "limited" success.

The tide changed when the Sea Turtle Conservancy hired Leo Martinez, a *belador* (turtle hunter), to help understand turtles and work with people who made their living hunting turtles. The Sea Turtle Conservancy shifted its focus to protect the rookery, the twenty-two-mile black sand beach where green turtles nested. Costa Rica established Tortuguero National Park in 1970. Initially, quotas were set on harvesting the turtles. Ecotourism grew, and the park now sees fifty thousand visitors annually. And harvesting of the turtles was banned.[24] Since 1970, the number of turtle nests has increased from 20,000 to over 180,000. Research assistants, many of whom are Latin American, are now learning the art of turtle stewardship.[25]

MARINE POLLUTION PREVENTION

In 1973, the International Marine Organization adopted the International Convention for the Prevention of Pollution from Ships. A 1978 amendment accounts for oil spills. Annex IV essentially prevents the dumping of raw sewage into the ocean. Annex V generally prohibits the dumping of garbage into the ocean and requires ports to provide garbage disposal facilities.[26] The convention was most recently amended in 2011 to promote energy efficiency in ships to reduce the release of carbon dioxide.

The US Marine Debris Research Production and Prevention Act of 2006, and its 2012 reauthorization amendments, align the United States with international efforts.[27] This law has clout. Princess Cruises and its parent company, Carnival, were fined a total of $60 million in April 2017 and June 2019 for dumping oil-contaminated waste and food-related plastics into the ocean.[28]

Over and over, countries bordering a given body of water have forged treaties to discourage pollution.

Scandinavia created the Convention on the Protection of the Marine Environment of the Baltic Sea Area. Known as the 1992 Helsinki Convention, it uses the "precautionary principle" to direct preventive measures, even when there is no proof that pollution has occurred. When harm occurs, the convention applies the "polluter-pays principle."[29]

Other conventions to protect the marine environment apply to the Black Sea (Bucharest, 1992), the Caribbean (Cartagena de Indias, 1983), the Caspian Sea (Teheran, 2006), the east and west coasts of Africa (four conventions and protocols, starting with Nairobi, 1985), the Mediterranean Sea (a dozen conventions and protocols, starting with Barcelona, 1976), the Northeast Atlantic (Paris, 1992), the North Sea (Bonn, 1983), the Pacific (ten conventions and protocols), and the Red Sea and the Gulf of Aden (Jeddah, 1982).[30] There is a pattern here.

Does this international progress in healing our oceans count as one Handprint or a multitude of Handprints? Up to you.

OCEAN GYRES

Worldwide, each year millions of tons of land-based plastic flow down rivers and into the oceans.[31] Cycloidal currents gradually concentrate this debris into five giant "gyres" in the Western Pacific, the Eastern Pacific, the North Atlantic, the South Atlantic, and the Indian Ocean. Horror stories abound, such as albatrosses feeding plastic pellets to their young and jellyfish trying unsuccessfully to eat plastic.

Informed by this reality, California became the first state to ban single-use plastic bags, in 2014. Other states to do so include Connecticut, Delaware, Hawaii, Maine, New York, Oregon, and Vermont. Cities with plastic bag bans include Boston, Chicago, Los Angeles, San Francisco, and Seattle. The District of Columbia requires a five-cent charge for plastic or paper bags used for groceries or alcohol.[32]

There has been pushback. Using language promoted by the American Legislative Exchange Council, fourteen states have *outlawed the banning of plastic bags*.[33]

Globally, some sixty countries have banned plastic bags, over twenty in Africa. Bangladesh was the first in Asia in 2002, not very long ago by historical standards. Most bans have exceptions. Enforcement may vary. The manufacture and export of plastic bags may be allowed.[34] That said, on October 2, 2019—the anniversary of Gandhi's birth—India banned both the manufacture and sale of single-use plastic bags, cups, plates, small bottles, and straws.[35]

How do you clean up a mess the size of a continent, running as much as a hundred feet deep?[36] We may have a way. Boyan Slat, while an engineering student at Delft University of Technology in the Netherlands, proposed using booms connected to floating platforms to collect the floating plastic. Everything happens in slow motion so fish can swim away or under the booms.[37] A 2014 crowdfunding campaign to scale up The Ocean Cleanup raised over $2 million from thirty-eight thousand people.[38]

In 2019, the first full-scale cleanup system, dubbed Wilson, was built in Alameda, California, and demonstrated in the Pacific Garbage Patch (that is, North Pacific Gyre). Segmented high-density polyethylene floaters extended up to 0.6 mile (1 kilometer) either side of a seagoing collection station. The Ocean Cleanup envisions scores of durable collection systems, serviced by pickup vessels, funded by large-scale recycling. Meanwhile, The Ocean Cleanup is developing a boutique market for its *verifiable* ocean plastic. The organization has also initiated efforts to capture plastic from rivers before it enters the ocean.[39]

Is the original Ocean Cleanup idea morphing? Maybe it needs to. It took more than nine years for Amazon to turn a profit, and they now sell a lot more than books.[40] Creating large-scale Handprints usually requires commitment *and* creativity.

HANDPRINT OPPORTUNITIES

17.1 Join a beach cleanup. Check out the International Coastal Cleanup interactive map on the Ocean Conservancy website.

17.2 Donate to, or work for, organizations that conserve the qualities we love about the ocean, such as these:

- **The Ocean Cleanup Foundation** develops advanced technologies to rid the oceans of plastic.
- **The Ocean Conservancy** creates science-based solutions for a healthy ocean and the wildlife and communities that depend on it. Their work ranges from coastal cleanups to researching ocean acidification.
- **Save the Manatee Club** focuses on West Indian manatees and their habitats off the coasts of Florida, Central America, and northern South America. Related to the elephant, manatees are plant-eating sea mammals weighing about half a ton.[41]

- **The Sea Shepherd Conservation Society** uses innovative direct action to stop the destruction of habitat and slaughter of ocean wildlife.
- **The Sea Turtle Conservancy** ensures the survival of sea turtles within the Caribbean, Atlantic, and Pacific through research, training, and advocacy.

17.3 Participate in an ocean citizen science project. The National Oceanic and Atmospheric Administration coordinates sixty-five projects that crowdsource citizen observations. Monitor whales on the California coast. Report marine debris anywhere in the world. Become a marine sanctuary steward for the Florida Keys or the Gulf of the Farallones (off the north and central California coast).[42]

17.4 Propose to ban plastic bags in your community. Check Green That Life for guidelines and resources.[43]

17.5 Ban the butt. Tom Novotny, epidemiologist at San Diego State University, proposes that laws be passed to stem the flow of plastic cigarette filters into the environment.[44]

17.6 Study marine biology. Understand and partner with Earth's cradle of life. Check out the long list of colleges provided by the MarineBio Conservation Society.[45]

IN YOUR JOURNAL

How will *you* become a steward of our oceans?

VII

WITHIN OUR GRASP

18

A CIRCULAR ECONOMY

Extract materials from the earth, craft them into things, briefly use the products, then dispose of them. We deplete the earth while landfills and oceans fill with garbage. That defines unsustainability.

Instead, we need an economy that does not depend on growth—a circular economy. The good news is that we have experience with valuable elements of such an economy. Resale. Recycling. Benefit corporations. Cradle-to-cradle design. However, barriers to a planet-friendly economy include negative attitudes toward used things, exclusively profit-centric business models, and the extraordinary rights given corporations. Handprint thinking builds on our experience and creatively responds to our challenges.

RESALE

The resale industry makes the second "R" in "reduce, reuse, recycle" real. Our clothing and household donations help fuel the resale industry. Our purchases keep it thriving. Many of us are familiar with giving our old furniture and outgrown clothing to Goodwill or other thrift stores. Value Village helps disabled people. Deseret Industries sends leftover clothing to developing countries. In 2018, Goodwill Industries diverted 4 billion pounds (1.8 billion kilograms) from landfills and helped 242,000 people train for employment in careers such as health care, information technology, and banking.[1] Meanwhile, I get great deals on a file cabinet, Hawaiian shirts, and handmade baskets.

The resale industry may be bigger than you imagine. There are twenty-five thousand resale, consignment, and antique shops in the United States. Their collective revenues total approximately $17.5 billion (2019).

All of those sales represent goods that did not need to be made from virgin resources.[2] My plaid cowboy shirt with snap buttons comes from Buffalo Exchange, which operates forty-five stores in seventeen states. Fashion-conscious Crossroads Trading has more than thirty resale locations.

While not formally counted as part of the resale industry, environmental dividends flow from the resale of cars, boats, motor homes, and pawned items. Eco-dividends also flow from resale via Craigslist, eBay, office liquidators, and auctions.

RECYCLING

Recycling is an established part of our circular-economy tool kit. And there is room for improvement.

Recology is a West Coast recycling and waste management company, owned by its three thousand employees. Recology Waste Zero specialists partnered with the Google sustainability team at Mountain View, California. Since 2011, they increased the campus waste "diversion" rate to recycling and composting from 37 percent to 68 percent. A waste audit and working with kitchen staff proved crucial to this turnaround. Perhaps as challenging, another Recology project increased the diversion of waste to recycling by 35 percent in a 324-unit SeaTac, Washington, apartment complex. The strategy there included multilingual fliers for residents and tours of a resource recovery facility for their children.[3]

According to the US Environmental Protection Agency, the municipal recycling and composting rate for the country was 35.2 percent in 2017. (Construction and demolition rates were calculated separately.)[4] In 2018, over two-thirds of the paper in the United States was recycled, nearly double the rate in 1990. It is relatively easy to recycle glass and metal; even so, the United States only recycles 26 and 30 percent, respectively.[5]

Recycling plastic is more challenging. The Association of Plastic Recyclers makes it their business to figure out how. There are more than twenty types of plastic and many hybrids. In 1988, the industry started putting resin identification codes on plastics to facilitate recycling. That helps, especially with more readily recycled #2 and #4 plastics. TerraCycle's customer-pays-for-the-service business model provides a way to collect and arrange recycling for most forms of plastic. However, a fundamental challenge is that recycled plastic costs about three times as much for manufacturers as virgin (fresh from fossil fuel) plastic. Pricing carbon (chapter 19) could change this calculus.[6]

Laws that ban single-use paper and plastic bags and put a deposit on beverage containers help curb our waste stream. In addition, eco-conscious consumers have at least three important points of leverage in this system: We can use and help improve existing recycling collection services. We can pay extra for recycled products. And we can avoid buying clothing, toys, and packaging made with unrecycled plastic. In sum, we can let go of the idea that recycling is someone else's problem.

THE BENEFIT CORPORATION

Benefit corporation status gives for-profit businesses legal permission to consider the social impact and the service of the organization primary over shareholder return. Otherwise, corporate directors are legally bound to maximize return to stockholders, setting the stage for environmental abuse. In 2010, Maryland became the first state to approve benefit corporation legislation. Over thirty-five states have now passed similar legislation.[7]

Benefit corporation status helps align the directors, employees, customers, and community. Suddenly, the reality of the soulless corporation shifts in favor of a circular economy.

B CORPORATIONS (B CORPS) VIS-À-VIS BENEFIT CORPORATIONS

B Labs administers a certified benefit corporation program that is available throughout the United States and internationally. The requirements for certification are much the same as for state-chartered benefit corporations, except that independent oversite by B Lab provides greater transparency. B Labs rates each company in five areas: environment, customers (who is served), community, workers, and governance. Companies can both charter with the state and certify with B Labs. When a company is sold, the buyers in most states may drop (or continue) the company's benefit corporation status.

Source: "FAQ," B Lab, accessed August 29, 2016, http://benefitcorp. net/faq.

Deltec Homes earns B certification points by building net zero energy homes out of Ashville, North Carolina.[8]

The W. S. Badger company makes Badger Balm, which works wonderfully to heal cracks in my heels. They serve their New Hampshire employees organic lunches and organize a biannual employee service day—in addition to sourcing organic ingredients and giving 10 percent of their pretax profits to charity. "At our core, we are a mission-driven business and our mission in its simplest form is 'healing products, healthy business, make a difference.'"[9]

Patagonia, based in California, helped pioneer the concept of a benefits-driven company. Patagonia sources 75 percent of its outdoor apparel from "environmentally preferred sources" such as recycling and organic suppliers. Suppliers are shown transparently on Patagonia's website. In 2018, Patagonia adopted a new mission statement: "We're in business to save our home planet."[10]

As of 2020, there are more than 3,500 benefit corporations and B-certified corporations in seventy-five countries.[11] Leather Heart makes stuffed toys out of recycled fabric in Caracas, Venezuela. BioCarbon Partners in Zambia provide forest carbon offsets. Watts Battery, located in Moscow, provides solar electricity plug-and-play modules. The Wave in Hong Kong provides coworking and event space with an emphasis on community development, environmental ethics, and cultural diversity.

CRADLE TO CRADLE

In the 1990s, Interface, a carpet manufacturer led by Ray Anderson, started looking for ways to be more sustainable. They developed the idea of modular carpet, which could be replaced square by square as it wore out. The returned carpet could then be made into more carpet. Tax laws and customer practices eventually forced a shift from the evergreen service agreement to a focus on a more sustainable supply chain and tapping into existing waste streams such as discarded fishing nets. Along the way, Interface's pioneering effort—and Anderson's witness to his peers in industry—sparked a revolution in industrial sustainability.[12]

Around the same time, in 1992, architect William McDonough and chemist Michael Braungart teamed up to publish *The Hannover Principles: Design for Sustainability*. In 2002, they followed up with *Cradle to Cradle: Remaking the Way We Make Things*. These books and two decades of consulting with industry provided the intellectual capital for the Cradle to Cradle Product Innovation Institute, founded in 2010.[13]

The Cradle to Cradle (C2C) Certified Product Standard addresses the entire product cycle, not just the moments when it is in the hands of the consumer. It attests to satisfying requirements in five categories: material health, renewable energy and carbon management, water stewardship, social fairness, and circular economy (material reutilization). That last category is further broken down into three areas: use of rapidly renewable or recycled materials; degree to which the materials in the product can be reused, recycled, or composted; and establishing whether the product can be safely returned to industry or nature.[14]

Duluth, Minnesota–based Loll Designs makes Cradle to Cradle Silver Certified outdoor furniture from recycled milk jugs. In the certification process, the company's consultant, MBDC, worked with suppliers to avoid toxic heavy metals and materials containing organohalogens. Loll accepts worn-out returns, which are recycled into new furniture.[15]

Playworld, in Lewistown, Pennsylvania, makes C2C-certified community playground systems. They use polyethylene-based Eco-Armor for their Silver-certified products. This eliminates nearly all PCBs and their toxic manufacturing and incineration by-products. Playworld tracks the carbon Footprints of their suppliers as well as themselves. Their mantra is "Without a healthy world where would we play?"[16]

Rajby Textiles in Pakistan met the requirements for the Platinum (highest) level of C2C certification for its Beluga denim. Worn-out fabric will be recycled into new products as part of the company's takeback system.[17]

Biofoam, manufactured by Synbra Technology also appears on the Cradle to Cradle Certified Products Registry. It is used for cavity wall house insulation, surfboards, and organ transport boxes. This product uses polylactic acid derived from renewable vegetable materials. Biofoam can be remolded into new products, and it is compostable in industrial (relatively high-temperature) facilities.[18]

While research is required to create these products, the science of industrial sustainability is replicable. Cradle to Cradle maintains a database on over thirty thousand chemicals. As of 2020, more than six hundred Cradle to Cradle certifications, applying to many times that number of products, have been granted to companies in thirty countries.[19]

WE THE PEOPLE

In the United States, money has been a driving factor in governance. Researchers for Princeton and Northwestern Universities analyzed nearly

1,800 opinion polls taken between 1971 and 2002. After the influence of organized groups and large-money interests were taken into account, researchers found that the *majority opinion had essentially no effect on policy outcome*, whether the public was for a policy or against it.[20] Prominent among those interests: large corporations.

Corporations in the United States were originally established as public service entities for a specific time period, like twenty years. Harvard University was the first US corporation. Then the US Supreme Court decided that the corporation charter was a contract that could not be broken by the state (*Dartmouth College v. Woodward* [1819]). A century later, the Supreme Court established that the primary purpose of a corporation was to provide profit to its stockholders (*Dodge v. Ford Motor Co.* [1919]).[21]

Over time, the Supreme Court has essentially given corporations the right of personhood. The Fourteenth Amendment of the Constitution (1868) reads in part, "nor shall any State deprive any person of life, liberty, or property without due process of law; nor deny to any person within its jurisdiction the equal protection of the laws." According to Thom Hartmann, author of *Unequal Protection: How Corporations Became "People"—and How You Can Fight Back*, a clerk mischaracterized a court decision in 1886, stating, "The defendant Corporations are persons within the intent of . . . the Fourteenth Amendment."[22] The error stuck. Three years later, in *Minneapolis and St. Louis Railroad Co. v. Beckworth* (1889), the court decision did include corporations in the definition of "person." In due course, corporations gained immunity from surprise inspections by the Occupational Safety and Health Administration (*Hale v. Henkel* [1906]). Corporate advertising, whether factual or not, formally received First Amendment protection under *VA. Pharmacy Board v. VA. Consumer Council* (1976). Likewise, campaign contributions were accorded the equivalence of free speech under *Buckley v. Valeo* (1976).[23] The right of corporations (and wealthy people) to make unlimited contributions to Political Action Committees was granted by the Supreme Court in the *Citizens United v. Federal Election Commission* in 2010.

Granting of effective corporate personhood, plus the interpretation of campaign contributions as free speech, gives corporations and wealthy people a profound ability to influence elections—and therefore environmental policy.

A proposed amendment to the constitution would deny artificial entities—corporations—constitutional rights that are accorded to persons. The intent is to give our government the ability to better regulate corporations in the public interest and, indirectly, in the interest of the environment. This amendment would also declare that money is *not* free speech. In other words, the people, through the legislative process, could limit the influence of corporate and individual wealth in making decisions that affect everyone.

In 2019, this proposal, known as House Joint Resolution 48, gained over sixty cosponsors in the US House of Representatives.[24] Enacting such

PROPOSED CONSTITUTIONAL AMENDMENT TO DENY CORPORATIONS CONSTITUTIONAL RIGHTS

Section 1.

The rights protected by the Constitution of the United States are the rights of natural persons only.

Artificial entities established by the laws of any State, the United States, or any foreign state shall have no rights under this Constitution and are subject to regulation by the People, through Federal, State, or local law.

The privileges of artificial entities shall be determined by the People, through Federal, State, or local law, and shall not be construed to be inherent or inalienable.

Section 2.

Federal, State, and local government shall regulate, limit, or prohibit contributions and expenditures, including a candidate's own contributions and expenditures, to ensure that all citizens, regardless of their economic status, have access to the political process, and that no person gains, as a result of their money, substantially more access or ability to influence in any way the election of any candidate for public office or any ballot measure.

Federal, State, and local government shall require that any permissible contributions and expenditures be publicly disclosed.

The judiciary shall not construe the spending of money to influence elections to be speech under the First Amendment.

Section 3.

Nothing in this amendment shall be construed to abridge freedom of the press.

Source: House Joint Resolution 48 introduced February 22, 2019. "H.J.Res. 48: Proposing an Amendment to the Constitution of the United States Providing That the Rights Extended by the Constitution Are the Rights of Natural Persons Only," GovTrack, accessed October 5, 2019, https://www.govtrack.us/congress/bills/116/hjres48/text/ih.

an amendment, which requires the concurrence of three-fourths of the states, would be a significant collective Handprint.

One of the basic functions of a corporation is to protect the entrepreneur from creditors if his or her venture does not work out. Nothing in the proposed constitutional amendment changes that attribute of our free-enterprise system. Instead, an unmanageable barrier to the creation of a circular economy would be removed.

HAPPINESS

Another hindrance to creating a circular economy is our semiofficial definition of well-being, the gross domestic product (GDP). Many worthy things, like raising children and services provided by nature, are not fully measured by the GDP. Conversely, the GDP rises when we eat unhealthy food and go to the hospital.

In 1972, Jigme Singye Wangchuck, the fourth king of Bhutan, declared that the gross national happiness was more important than the gross national product. What can this small Himalayan nation between India and China teach us? How can happiness be measured? How can it be used to govern?

In 2010, after years of analysis and data collection, Bhutan formally adopted the Gross National Happiness Index of national well-being.[25] It has nine "domains": (1) psychological well-being, (2) health, (3) education, (4) culture, (5) time use, (6) good governance, (7) community vitality, (8) ecological diversity and resilience, and (9) living standards. Here we see that the well-being of the environment is analytically linked to the well-being of the nation.

Within the ecological diversity and resilience domain, there are four subtopics: pollution, environmental responsibility (focusing on eco-friendly attitudes), wildlife, and urban issues. Surveys of rural citizens assess concerns relating to wildlife's impact on agriculture. Urban dwellers report the impact and needs (such as green spaces) associated with rapid urbanization. Each environmental metric is tracked to discern whether it erodes as the country develops.

What does *The Economist* have to say about Bhutan? "The hydropower sector will continue to be the most important driver of economic growth [GDP], which is expected to average 7.6% in 2019–20. Electricity exports to India from the newly commissioned Mangdechhu project [dam] will support export earnings."[26] Hmm, hydropower is fueling their boom economy—sustainable in terms of global warming, but at what cost to the local environment? Also, there is no mention of gross national happiness. Bhutan's journey is a work in progress, as is our ability to embrace it.

Still, others are also picking up happiness as a relevant measure of progress. In 2007, Thailand offered their Green and Happiness Index. In 2011, the UN General Assembly passed resolution 65/309, "Happiness, Towards a Holistic Approach to Development."[27] Since 2012, the United Nations Sustainable Development Solutions Network has published annual World Happiness Reports, assessing 156 countries. Example finding: the provision of free long-distance calls in 2014 in Mexico peaked national happiness.[28]

In 2011, Seattle, Washington, launched their Happiness Initiative. Surveyors and analysts found that building community and improving time balance (including reducing commutes) are needed to improve happiness; happiness improves very little once annual family income rises above $75,000; and, youth between nineteen and twenty-four were the most disaffected. This is information a city council can use.[29] Unfortunately, Seattle's foray into happiness measurement did not account for the environment.

HANDPRINT OPPORTUNITIES

18.1 Buy and sell clothes and furniture from a resale store. Brag about it. Check out Value Village and Goodwill nationwide. Check out Free Geek for reconditioned computers in Portland, Oregon. Check out eBay for anything anywhere.

18.2 Rent instead of buying tools and cars. Rent-A-Wreck rents reliable used cars from Bakersfield, California, to Stockholm, Sweden.

18.3 Buy goods made from recycled materials, including clothing, toys, and office supplies.

18.4 Recycle paper, metal, glass, and acceptable plastics through recycling programs.

18.5 Use TerraCycle Zero Waste Boxes to recycle and upcycle anything from your bedroom, your event, or your worksite. TerraCycle charges a fee, but shipping is prepaid. A Dining Disposables and Party Supplies Zero Waste Box goes for a little over a hundred dollars.[30]

18.6 Buy goods and services from benefit corporations (or B Corps). Check out the directory on the B Lab website.

18.7 Buy products with Cradle to Cradle Certified materials, like Seaweed Insulation, BioFoam, and Mushroom Materials (instead of Styrofoam), and vinatur outdoor fabrics. Check out the Cradle to Cradle Products Registry.[31]

18.8 Insist that businesses you frequent publish an annual sustainability report. The act of writing the report establishes a motivating benchmark for improvement.

18.9 Support an amendment to the Constitution to end corporate personhood and end money as a form of free speech. Check out Move to Amend. Check out the Brennan Center for Justice at New York University School of Law to track legal challenges to corporate citizenship. Read *Unequal Protection: How Corporations Became "People"—and How You Can Fight Back* by Thom Hartmann.

18.10 Value happiness in personal and national decision making. Go for a happiness walk. Host a dinner around March 20, International Happiness Day. Check out Gross National Happiness USA. On their website, sign the Charter for Happiness, which calls for "economic, environmental, and societal justice by listening to the voices of the people, measuring what matters, and instituting practices and policies that support the conditions of happiness."

IN YOUR JOURNAL

How will *you* help evolve a circular economy?

19

THE PARIS AGREEMENT

Demonstrations around the world prefaced the 2015 Paris Conference on Climate Change. Between 2009 and 2014, the Athabasca River-keepers, along with First Nations partners, organized tar sands healing walks in northern Alberta to call attention to the lunar landscape left by bitumen mining. I was there in 2013. In September 2014, the People's Climate March consisted of more than two thousand events with over 1,500 organizations in 162 countries, culminating with a 400,000-person event in New York City.[1]

A parallel effort within the collective governance of our planet also proceeded at its own pace.

THE UN FRAMEWORK
CONVENTION ON CLIMATE CHANGE

The groundwork for world cooperation on climate change was laid with the United Nations Framework Convention on Climate Change (UNFCCC). This accord established an ultimate objective to "stabilize greenhouse gases concentrations in the atmosphere at a level that will prevent dangerous anthropogenic (human-induced) interference with the climate system."[2] The UNFCCC places more responsibility for emissions reduction and financing on developed countries and establishes the importance of cooperation with the forty-nine least developed countries. The framework establishes a general approach for gathering data about a given country's progress in reducing greenhouse gas (GHG) emissions against 1990 benchmark levels. Delegates finalized the UNFCCC at the 1992 Earth Summit held in Rio de Janeiro, and it entered into force in 1994; 197 parties, including the United

States, are now committed to this agreement. For more than a quarter century, the UNFCCC has served as a platform for the halting but discernible progress in our planet's encounter with climate change.

The 1997 Kyoto Protocol entered into force in 2005. It established legally binding GHG reduction obligations for industrialized countries and the "economies in transition" of eastern Europe and the former Soviet Union. Action plans addressed two periods: 2008–2012 and 2013–2020. Afghanistan, Sudan, and the United States failed to sign the Kyoto Protocol. However, 192 parties, including Russia, China, and the European Union, *did* ratify it. While emission reductions were reported under Kyoto, the main benefit was probably the experience gained in collaborating on the minutia of climate science.[3]

The Copenhagen Accord of 2009 established two degrees Celsius above preindustrial levels as the globally agreed temperature rise limit. While it disappointed activists who wanted an agreement to bound our collective behavior, the Copenhagen Accord served as a catalyst for follow-on work.[4]

The Cancun (Mexico) Agreement of 2010 established the Climate Technology Centre and Network (CTCN) to respond to requests from the forty-eight least developed countries. The CTCN became fully operational in 2013. Funded by developed nations, including the United States, a network of consultants and institutions provide expertise.[5] The stream of requests for CTCN help include accelerating Panama City's transition to low-carbon mobility, improving the capacity for recycling waste and organic material in Gambia, and developing saline water purification technology at the household level in Bangladesh.

At the 2011 Durban, South Africa, Conference of the Parties (COP) to the Kyoto Protocol, world leaders wrestled with their continuing inability to strike another binding agreement. The meeting ran two days longer than planned—until participants agreed to convene a substantive Paris meeting.

Two years later, COP negotiators in Warsaw agreed to "mobilize" $100 billion—per year—for the Green Climate Fund to help implement GHG reduction in low-income and African countries and small island nations. Participants also agreed to help compensate countries for *not* cutting their forests. In addition, a $100 million adaption fund was capitalized (read: ready to spend) to help implement projects that mitigate climate change impacts.[6]

Supporting this process, the Intergovernmental Panel on Climate Change (IPCC) provides policy makers with ongoing scientific assessments of the physical basis for climate change, the adaptations that are needed, and the mitigations that are available. The IPCC shared the 2007 Nobel Peace Prize with Al Gore "for their efforts to build up and disseminate greater knowledge about man-made climate change, and to lay the foundations for the measures that are needed to counteract such change."[7] While the IPCC marshaled scientific evidence, Gore's movie *An Inconvenient Truth* and its sequel made a graphic case to the public.

THE PARIS AGREEMENT

Officially, the meeting in Paris from November 30 to December 12, 2015, was the twenty-first COP (Conference of the Parties) of the UNFCCC and the eleventh session of the Meeting of the Parties of the Kyoto Protocol.

At the outset of the Paris conference, Avaaz organized a silent demonstration of thousands of shoes in Place de la République. Even the secretary general of the United Nations and the Pope sent shoes. Early in the proceedings, demonstrators stood in a long line, holding red cloth in front of themselves, drawing attention to the needs of the poor island nations and indigenous people. Their "red line" called for climate justice and urged delegates to "leave it [carbon] in the ground." During the conference negotiations, a kayak action on the Seine River called attention to the Earth Charter (chapter 6) and protested that references to indigenous people around the world in the budding agreement were not binding.[8] (The Kyoto Accord, eighteen years earlier, failed to mention indigenous people at all.)

Collectively, these actions, and many more, created a political climate in which high aspirations (if not the highest) could prevail.[9] On December 12, 2015, 195 nations and the European Union unanimously approved the Paris Agreement.[10]

KEY PROVISIONS OF
THE PARIS AGREEMENT OF 2015

Shared aspiration. The world now aspires to keep the average global temperature below 1.5 degrees Celsius above preindustrial levels. Note, we just hit 1.0 degree Celsius in 2015. The old (Copenhagen 2009) target of 2.0 degrees Celsius is still in the document but carries no more weight than the "high ambition" goal.

National intentions. Parties are expected to submit "intended nationally determined contributions" to minimize greenhouse gas emissions and provide sinks for carbon. As of the conference, 187 countries submitted pledges.

Scientific assessment. Scientists are invited to advise the proceedings. Scientists tell us that we are currently trending toward 4.0 degrees Celsius above preindustrial levels. National intentions submitted as of the Paris proceedings trim that projection to a 2.7- to 3.0-degree Celsius rise in the average global temperature—still a long way from 1.5.

Money for developing countries. Developed countries promise (grimace if you must) that $100 billion every year will flow to developing countries to support both their efforts to deal with climate change and their need to eradicate poverty.

Financial accountability. Developing countries receiving the money are accountable through results-based payments, transparency, and reporting that is consistent with international standards.

Accommodation. It is acknowledged that nations with limited infrastructure will need accommodation to meet reporting requirements.

Adaptation. It is acknowledged that there will be a need for some nations, especially island nations, to adapt to the effects of global warming.

Five-year cycle. A "global stocktake" will be performed every five years, starting in 2023. Assessments will be made to set the stage for adjusting commitments accordingly.

Toward carbon neutrality. By the second half of the century, each nation will move toward a balance between greenhouse gas emissions and the capacity of greenhouse gas sinks. That is, we seek to be carbon neutral as a world. Note, at least six nations are already net sinks of carbon.

Kyoto Protocol. Commitments most of the world (not including the United States) made in meeting the Kyoto Protocol are still relevant. Parties of the Kyoto Protocol are asked to complete their work and ramp up to full implementation of the Paris Agreement by 2020.

United Nations. A whole cast of United Nations institutions are marshaled to ensure that we are successful in dealing with climate change.

Source: "Adoption of the Paris Agreement," United Nation Framework on Climate Change, adopted December 12, 2015, http:// unfccc.int/resource/docs/2015/cop21/eng/l09.pdf.

The Paris Agreement is not completely binding, but it is functional. Diplomats gave us their best effort, given that India was not going to let go of coal and the United States could not deny its own conservative constituency. The agreement could not have been written twenty years earlier. Our science has advanced. We build on a raft of previous agreements. The arcane language of international treaty making is understood by a wide cast of characters. Even the procedural art of reaching such a complex agreement has been honed over time.

The Paris Agreement also sidestepped some potential showstoppers. President Obama was able to approve the Paris Agreement through executive order rather than Senate ratification. The Paris Agreement elaborates on the UNFCCC, which *was* approved by the US Senate. Article 28 requires a party to wait at least four years before withdrawing from the agreement.[11]

Developed nations provided funds to avoid the legal quagmire of formally pleading guilty for reducing the capacity of our biosphere to maintain a stable environment.

Fossil fuel companies are not required to part with their mega fortunes. No one is demonized. Renewables rather than bureaucrats will need to rise to that challenge (chapter 14). Likewise, divesting, modest living, and a change in the way we do business will likely be needed.

Negotiators did not choose a preferred technology. Large environment-stressing dams might be brought online, as could radioactive-waste-producing nuclear power plants. Options for dangerous large-scale interventions in the earth's natural systems (geoengineering) are still open.[12] Dialogue and protest on these fronts will continue.

The Paris Agreement avoids any heroic assumptions that we will miraculously invent a psychology or technology to solve global warming. But it invites us to do so.

Heroes abound. Christina Figueres of Costa Rica ensured that every nation was ready to constructively participate in the Paris proceedings.[13] French foreign minister Laurent Fabius presided over the shuttle diplomacy needed to reach the final agreement. Tony deBrum, the Marshall Islands foreign minister, formed the High Ambition Coalition that moved the emerging agreement from a take-it-or-leave-it ultimatum to a tour de force of statesmanship.[14]

The Paris Agreement acknowledges that some countries (read: India and China) have yet to reach a peak in their GHGs. That is a fierce challenge. India's prime minister Narendra Modi's announcement of a 120-nation solar initiative, seeded with $30 million, may not be sufficient to offset an increase in coal production.[15]

There is plenty to criticize. The agreement inadequately addresses concerns of indigenous people. It fails to mandate carbon pricing.[16] Greenhouse gas reduction commitments are voluntary. So are contributions to nations with limited financial and technical capacity. "This deal offers a frayed life-line to the world's poorest and most vulnerable people," said Helen Szoke of Oxfam.[17] Paul Cook of Tearfund says, "What has been exciting is to see the growing movement these talks have fostered."[18] Bill McKibben, of 350.org, looked at the road ahead. "Since the pace is the crucial question now, activists must redouble our efforts to weaken [the fossil fuel] industry."[19]

Other activists are more positive. "This deal will represent a turning point in history, paving the way for the shift to 100 percent clean energy," said Emma Ruby-Sachs of Avaaz.[20] "The agreement's temperature goal, net zero emissions objective, and processes to steadily increase the ambition of national emissions reduction commitments combine to send a clear message to the fossil fuel industry . . . your efforts to block action on climate change are no longer working," said Alden Meyer, director of the Union of Concerned Scientists.[21]

Jim Yong Kim, president of the World Bank, says that the Paris Agreement will "trigger the flow of money to a carbon neutral world. There is no development without tracking climate change."[22] In other words, we are unleashing powerful market forces in service of the common good—potentially key to meeting our 1.5-degrees-Celsius goal.

My opinion: We created a world-class tool—a Handprint in its own right—for meeting the challenges of climate change. If properly maintained, it will serve our grandchildren well. I count December 12, 2015, to be a significant date in history. However, Paris is no more the end state of climate change conferences than the Constitution is the end point of the US government.

The Paris Agreement entered into force on November 4, 2016, after fifty-five nations (including the United States and China) had registered formal acceptance. This was well ahead of the timeline for full implementation by 2020. By the end of the Marrakech, Morocco, Climate Change Conference, two years later, 110 nations were officially on board.[23] And the Climate Vulnerable Forum, consisting of 48 nations, committed to 100 percent renewable energy by 2050.[24]

PUTTING A PRICE ON CARBON

In a widely reported speech to Congress in 1988, Dr. James Hansen, then director of NASA's Goddard Institute for Space Studies, warned the world about global warming. After the Paris Agreement was signed, he asserted that we must put a price on carbon to reduce overall emissions.[25] Money is a language that businesses understand. Adding a system feedback loop can have a profoundly transforming effect on the market.

From a public policy perspective, the price of carbon represents the damage additional carbon in the atmosphere does—the hurricanes, the droughts, the resulting cultural instabilities. The US Environmental Protection Agency calculated a "social cost" of carbon in the atmosphere of thirty-seven dollars per ton (forty-one dollars per metric ton). Frances Moore and Delavane Diaz, a Stanford University team, took into account long-term economic effects of climate change to calculate the social cost at $220 per ton ($243 per metric ton).[26] Carbon pricing takes one of two explicit forms: a fee or tax on carbon-based fuels and a carbon trading system.

A fee may be added to the price of auto gasoline, as California does. We have long paid for gasoline taxes. A dime added to the price of a gallon of gas amounts to $15 per ton ($16.50 per metric ton) of carbon. However, using gasoline taxes to pave highways increases gasoline consumption—unless, perhaps, they are coupled with a transition to electric mobility. New Zealand imposes a carbon fee on forestry and even municipal waste.[27]

Carbon trading ("cap and trade") limits the amount of carbon released and sets up a market for the authorized emissions. Finland implemented one of the world's first carbon cap-and-trade programs in 1990. That presaged the European Union's adoption of carbon trading in 2005, affecting 45 percent of its carbon emissions with a cap that declines annually.

In 2008, the United Kingdom passed the Climate Change Act, which set a legally binding target of cutting carbon emissions by 80 percent by 2050. To meet this requirement, the UK established a carbon tax in 2013. This made coal more expensive than natural gas as a fuel for power plants. From 2012 to 2017—six years—coal dropped from supplying more than 40 percent of the UK's electricity to just 7 percent. During that same period, the price of renewable energy was falling. Renewables moved from supplying 4 percent to almost a third of the UK's electricity.[28]

A government's political process determines how income generated from carbon pricing is used. British Columbia and Ireland use it to lower taxes. California uses it to help low-income people. Costa Rica pays landowners to grow trees. The European Union sets aside some of its cap-and-trade allowances to fund "innovative renewable energy technologies and carbon capture and storage."[29]

New England and California pioneered carbon pricing in the United States. The Regional Greenhouse Gas Initiative focuses on power generation in eleven states: Connecticut, Delaware, Maine, Maryland, Massachusetts, New Hampshire, New Jersey, New York, Rhode Island, Vermont, and Virginia.[30] California's economy-wide cap-and-trade system, initiated in 2013, engages industrial emitters and fuel suppliers as well as electric generation.[31] Nationally, carbon dioxide emissions from power plants can be curtailed under provisions of the Clean Air Act of 1970, but executive branch (presidential) follow-through is needed.[32]

The World Bank 2020 report on carbon pricing reports "61 carbon pricing initiatives in place or scheduled for implementation." These are about evenly divided between emission trading systems (EMSs) and carbon taxes. Highlights include New Brunswick's carbon tax in Canada, deployed EMSs in Germany and Virginia, and a pilot EMS in Mexico—the first in Latin America. Reduced fossil fuel use during COVID-19 resulted in delayed implementation of emission trading in some cases. Still, as of 2020, carbon pricing covers 22 percent of the world's greenhouse gasses, up from 20 percent in 2019.[33]

Carbon pricing is not limited to the public sphere. CDP Worldwide, headquartered in London, keeps track of carbon tracking initiatives within companies. Some 150 companies around the world have instituted internal ("shadow") carbon pricing in their accounting to encourage internal movement toward greenhouse gas reduction. These include Exxon-Mobil, Dow Chemical, and Bank of America.[34] This is not greenwashing; these companies are preparing for the changing economic environment.

CITIES (AND EVERYONE ELSE) VERSUS CLIMATE CHANGE

On June 1, 2017, US president Donald Trump announced plans to withdraw from the Paris Agreement. In response, other parties to the agreement reaffirmed their support of the agreement, including India, China, Russia, and the European Union. Perhaps as important, the proactive response of cities may fill the vacuum left by national leadership. "President Trump may be withdrawing the U.S. from the Paris Accord," said Santa Barbara, California, mayor Helene Schneider. "But cities are stepping up and recommitting to adopt, honor and uphold the Paris climate goals."[35]

Cities are strategically positioned to make a powerful response to climate change. About 75 percent of energy-related global carbon dioxide emissions can be linked to cities.[36] Even when national governments get bogged down, cities can develop policies based on the pioneering work of other cities. In 2015, San Diego, California, became the first US city to commit to 100 percent renewable energy—by 2035.[37] The 2018 update of the Green Cincinnati (Ohio) Plan includes eighty strategies to reduce its carbon emissions by 80 percent by 2050. This includes the country's largest city-owned solar array.[38]

The Global Covenant of Mayors (GCoM) 2018 report "Implementing Climate Ambition" states, "Of the 9100+ GCoM committed cities, over 6000 have already established a strategic climate action plan for their communities defining their approach to meet their commitment. Collectively through overall commitment to the GCoM initiative, reported data show that 1,818 cities have reduced emissions by 20% (or 0.43 Gt) from their highest points of reported emissions."[39]

A key supporting partner of GCoM is C40 Cities (now more than ninety cities), which represents more than 700 million people, an eighth of the world population, and a quarter of its monetary economy (GDP). C40 highlights on-the-ground initiatives that will help us meet our aspirational temperature limit. Houston is turning a methane-leaking landfill into the Sunnyside Energy Project, implementing community solar generation for residents and businesses.[40] Chicago's Garfield Green is a net zero carbon and net zero energy housing project, built with replicable modules made at a local factory.[41] Addis Ababa, Ethiopia, is building light rail transit.

The United Kingdom cooperates with twenty-five other countries to adopt low-emission buses.

Copenhagen, Denmark, uses its leadership in urban innovation as a strategic business asset to foster "sustainable urbanization." In 2016, a coworking space called BLOXHUB opened its doors to an admixture of member companies (including start-ups), organizations, research institutions, and municipalities from both Denmark and around the world. Urbantech, a recurring BLOXHUB initiative, brings together start-ups and established Danish companies in a three-month accelerator program. Technologies represented in the 2020 cohort include local air quality sensing and analysis, a digital platform for climate-neutral building designs, and smart windows that regulate indoor climate while reducing energy use.[42]

Also with a sustainable city objective, the City Energy Project maintains a resource library of energy code best practices with an emphasis on existing buildings (see chapter 14).

We are also reaching a critical mass of states that have made clean power commitments. On September 10, 2018, Governor Jerry Brown signed SB 100 into law—California's commitment to 100 percent clean energy by 2045. "When it comes to fighting climate change and reducing our reliance on fossil fuels," said Kevin de León, the bill's chief sponsor, "California won't back down."[43] This intention is backed up by interim commitments of 50 percent renewable energy by 2026 and 60 percent renewables by 2030. New Mexico, which previously generated half its energy from coal, committed to 100 percent energy "from carbon-free sources" by 2045.[44] On July 18, 2019, Governor Andrew Cuomo, with Al Gore looking on, signed legislation committing New York to 70 percent renewable energy by 2030.[45]

In 2017, the US Environmental Protection Agency reported that 880 Green Power Partners of all sorts had committed to 100 percent renewable energy. These include Apple, eBay, Google, Microsoft, Bank of America, General Motors, IKEA, Nike, Visa, and Walmart.[46] At least forty colleges and universities *already* obtain 100 percent of their electricity from renewable resources, and many schools are tackling the tougher challenge of eliminating their use of fossil fuels. New York's Colgate University is switching to all-electric vehicles, solar-heated hot water, and geothermal heating.[47]

Fulfilling America's Pledge, a 2018 report by the University of Maryland and the Rocky Mountain Institute, showed how planned actions by states, cities, and businesses will achieve two-thirds of the US 2025 carbon emission reduction goal that was announced in Paris, even without federal leadership.[48]

HANDPRINT OPPORTUNITIES

19.1 Advocate 100 percent renewable energy for your business, university, city, and country.

19.2 Create climate neutral action plans for you, your family, and your business. What combination of Handprints and reduced Footprints feels right to reduce your net carbon release to zero? Identify steps you will take. Set dates. Share your intentions with friends and associates.

19.3 Put a price on carbon. Support Our Climate, which fosters state initiatives such as NW Renews, Energize Rhode Island, Put a Price on It DC, and Renew Oregon. Also, voice your support of the Energy Innovation and Carbon Dividend Act in Congress along with the Citizens' Climate Lobby.[49]

19.4 Present the "Truth in 10" slideshow to family, friends, and civic groups. Nobel Prize winner Al Gore prepared it for our free use.[50]

19.5 Teach the fundamentals of the Paris Agreement to children. This is multigenerational work. *Our goal: no more than 1.5 degrees Celsius global average temperature rise above preindustrial levels. Our strategy: voluntarily commit to become climate neutral by 2050 plus periodically report progress in fulfilling that commitment plus help low-income nations do the same.*

19.6 Ensure that the United States fully recommits to the Paris Agreement. Check out Union of Concerned Scientists for relevant campaigns and actions.[51]

19.7 Witness and report meetings of the Conference of the Parties to the United Nation Framework Convention on Climate Change. Caring matters.

19.8 Find employment in sustainable development or climate change mitigation. Check out companies and organizations mentioned in this book (see index), and those participating in the Climate Technology Centre and Network.

IN YOUR JOURNAL

What combination of reducing Footprints and creating Handprints comprises *your* climate neutral plan?

20

BE A HERO IN YOUR OWN STORY

Just before I left for a sabbatical to travel around the country in 1989, Carla Perry, a writer friend of mine, left me with these words: "Be a hero in your own story." At the time, I was thinking in terms of climbing mountains. Indeed, I climbed Devil's Tower in Wyoming, Mount Washington in New Hampshire, and Guadalupe Peak in Texas. After six months on the road, I came back to my work in energy efficiency with Bonneville Power Administration with renewed enthusiasm. Over the next four years, I managed a program that resulted in the birth of wind farms in the Northwest, and I initiated plans to convene the nation's first industrial energy-efficiency conference. I am proud of these Handprints.

You too will encounter very real opportunities to be a hero in your own story and in the planet's story.

TRANSITIONS

Almost any change holds within it a potential for sustainable thinking. Many Handprint projects begin at a time of transition.

Brian Wagner moved from Milwaukee to Portland to continue his education. The change of location—and Portland's great transit system— prompted him to let go of his car. A change in lifestyle that reduces a pattern of consumption is also a Handprint in my book. With his education, Brian took a job with Beneficial State Bank, a benefit corporation. He also joined the local EcoVoices Toastmasters group to enhance his ability to make a difference when speaking. In the process, Brian became the group's president and doubled its membership.

Jeanne and Dick Roy went on a sabbatical before deciding to dedicate their lives and livelihood to sustainability. That meant that Dick would leave a successful legal practice, along with its income. They have since founded the Northwest Earth Institute (now Ecochallenge.org) and the Center for Earth Leadership.

Peter Kalmus and his wife launched his eco-friendly lifestyle when moving to Pasadena, California, to work at Jet Propulsion Laboratories on the Space Program. He now bikes to work and forgoes flying, cutting his carbon Footprint by 90 percent. The family uses a "humanure" composting toilet. Peter wrote about his experience in *Being the Change: Live Well and Spark a Climate Revolution*, which, along with an associated movie, is a Handprint in its own right.

Jaime Ott used a two-year stint in Zambia with the Peace Corps to better define, in her words, "what would mean something to me." The answer for her turned out to be *directly helping people in agriculture*. Toward that end, she returned to study for a PhD in plant pathology with the view of becoming an agricultural extension agent. Does this involve regenerative agriculture? Ott says, "I get bogged down with thinking that there is only one way to do it, whether organic or regenerative. There are lots of pieces that can be fit together." She observed that, like everyone, farmers don't want pesticides leaching into the groundwater. Her laboratory's research at the University of California Davis relates to *Phytophthora* pathogens (a genus of water mold) that affect almond and walnut trees, both perennial crops that are part of the regenerative farming portfolio.

Once we recognize transitions as invitations to act on our sustainable values, we can use even modest events to make changes. I invited neighbors I barely knew to my birthday party and gave away food, books, and nice things I no longer needed. While visiting, they saw and talked about the "What's Green About This Room?" signs posted around the house.

A significant transition occurs for many of us between the ages of thirty-five and forty-five. We experience a sense of disillusionment. Maybe it is coupled with a health crisis, losing a job, a divorce, or all of the above. We may experience anger or depression. The journey that I shared at the outset of this chapter was my response to the breakup of a relationship at age thirty-nine.

Barbara Bradley Hagarty, author of *Life Reimagined*, reports Danish research revealing that "a sense of purpose" far outweighs any other factor in determining employee satisfaction. Hagarty offers five suggestions to transition to a life with a sense of purpose at its center.[1] No need to limit ourselves to the workplace with Handprint opportunities all around us.

MAKING A CAREER TRANSITION, FIVE SUGGESTIONS PROPOSED BY BARBARA BRADLEY HAGERTY

- **Address your concerns directly.** If paying the mortgage is an issue, move to a smaller, more sustainable house. The *Living Big in a Tiny House* video channel is full of stories of people who happily meet their needs for shelter with low incomes.[a]
- **Dip your toe in the water.** Volunteering for a nonprofit is a proven way to add experience to your résumé and connections to your network and feel an immediate sense of satisfaction.
- **Make a relatively modest leap.** While on my sabbatical, I asked to be relieved of one of my projects. Thus, I became available, upon my return, to manage a new program.
- **Respect your innate traits and talents.** Rachel Carson loved literature even though she chose a career in marine science. In her off-duty hours, she wrote *The Sea Around Us*, a best seller in 1951. That gave her the financial means to quit her job with US Fish and Wildlife Service in 1952 and write full time. Her book *Silent Spring*, published in 1962, focused the public's attention on the indiscriminate use of pesticides and led to the banning of DDT.[b]
- **Start plotting sooner rather than later.** Both you and the world will be better off when you are fulfilling your destiny. Opportunities can come along unexpectedly.

Source: Barbara Bradley Hagerty, "Quit Your Job," *The Atlantic*, April 2016.

---

[a] Bryce Langston, "Living Big in a Tiny House," YouTube, *Living Big in a Tiny House*, accessed January 21, 2020, https://www.youtube.com/user/livingbigtinyhouse/videos?app=desktop.

[b] Wikipedia, s.v. "Rachel Carson," accessed August 13, 2019, https://en.wikipedia.org/wiki/Rachel_Carson.

Retirement is a particularly opportune form of transition. Except that "retirement" has at its root the connotations of going away, retreating, and going to sleep. No wonder so many people die soon after they retire. You will likely live longer (seriously) if you treat your formal retirement as a graduation.

When Melanie Plaut retired from a thirty-six-year career as a medical doctor specializing in obstetrics, she honored her "future orientation" and started work for the Portland chapter of 350.org. Her experience and innate aptitude prepared her well to join the Green New Deal Team and lead the Fossil Fuel Resistance Team.[2]

When I was first offered a voluntary early retirement, it made no sense to me. I was paid well to work in a worthwhile field, energy conservation. My employer and my boss treated me well. I enjoyed going to work in the morning. But many of my ideas did not fit within my employer's box. Maybe it was time to leave that box, nice as it was, to explore my broader interest in sustainability. I have never regretted my decision—*to graduate* from my old box. Otherwise, this book would likely not have happened.

CLAIMING YOUR HANDPRINTS

A year or two after I came upon the Handprint idea, a friend asked, "So, what do you do with it?" My answer was a Handprint workshop to help people follow through on their intentions to make the world a better place. The idea is quite simple. All that is needed is blank paper and drawing tools, the more colorful the better. It is in two parts, past and future. Give it a try.

Start by claiming the ecological good *you* have *already* done, your past Handprint. Draw an outline of your left hand on a blank sheet of paper. Let the five fingers represent things you have already done for the environment. Modest things like recycling and signing a petition are fine. If you need prompts, think in terms of (1) a time you have given money or food to an environmental cause, (2) a time you have volunteered for such a cause, (3) something you have done by yourself, (4) things you have done as part of a larger effort, and (5) something worthwhile you unsuccessfully tried to do. The experienced gained is valuable.

My friend Kenneth is a pet-sitter and Vietnam veteran. His past Handprint includes living in an artist community for a year, living without a car for ten years, being a vegan for twenty-five years, and strongly advocating that I include food sustainability in this book. He tried living in an RV, but the winter cold was depressing. He needed to find other ways to live his values.

So will you.

Now create your future Handprint. Outline your right hand on another piece of paper. For each finger, name one specific way you *want* to help heal our planet. Choose from the Handprint opportunities at the end of each chapter, notes in your journal, or your own aspirations. As a guideline: (1) Choose a onetime action. (2) Volunteer for an environmentally proactive cause. (3) Donate to an organization that is creating sustainability. (4) Engage your circle of influence—your family, colleagues, or community. (5) Invest in a sustainability project or a mission to change the system.

It helps if you share your future Handprint with a friend or a fellow workshop participant, sign it, and date it. Treating your Handprint as something important will help you follow through with your intentions.

My future Handprint, dated November 12, 2019, includes giving an Earth Day talk to my congregation, helping organize the Portland chapter of the Pachamama Alliance, donating to Columbia Riverkeeper, motivating groups I participate in to switch to reusable plates and utensils at potlucks, and publishing *Our Environmental Handprints*.

Treat your future Handprint not as a to-do list but as a calling. Let go of items and add items as you feel called to do so. Revisit it. Let it evolve. It is your heart speaking. The better you listen, the happier you and the planet will be.

WHEN SOMETHING BREAKS

Intentions, plans, and day-to-day follow-through prepare us for what may be our more impressive breakthroughs on the road to sustainability. You may have heard that the Chinese use the same character for *crisis* and *opportunity*.

In 2007, I embarked on a journey from our Oregon home to scatter my mother's ashes in the mountains of central Arizona. I traveled in our twenty-three-year-old Volkswagen Vanagon. In California's Mojave Desert, the engine suddenly lost power. I putt-putted it to the side of the road and waited until a passing semi called the state police, who called for a tow to the town of Blythe on the Arizona border. There, two auto shops told me our old friend was not worth fixing. I called home to tell Willow the sad news. Instead of a lecture about taking an old vehicle on the road, Willow said, "I've been wanting to try living without a car."

After completing my mission in a rental car, I came home on the bus. In true engineering fashion, I calculated how much money we could save by not owning a vehicle. I considered gas, maintenance, insurance, and depreciation (decreasing value over time). I also assumed we would use a taxi in emergencies, rent a car for a vacation, and give gas money to those who give us rides. That yielded $1,000 per year savings—and much more if we had wanted a rapidly depreciating newer vehicle. But we did not go car-free for thirteen years to save money. I also calculated that not parking a car in our driveway would reduce our annual carbon dioxide emissions by 7.3 tons (6.6 metric tons). A breakdown on the road catalyzed a quantum reduction of our ongoing Footprint!

Larger-scale disasters can bear environmentally valuable fruit as well.

In 1976, reporters David Pollack and David Russell tested several sump pumps in the Love Canal neighborhood of Niagara Falls, New York. They found toxic chemicals. A year later, reporter Michael Brown conducted a door-to-door survey and discovered there was an abnormal number of children with birth defects in the neighborhood. Subsequently, the New York State Health Department identified an unusual number of miscarriages.[3]

The story of Love Canal is now well known. A chemical company used the canal as a dump site. Relentless reporting and activism, especially by Michael Brown and Lois Gibbs (a local mother and homeowner) eventually moved the nation. Eight hundred families were relocated; they received compensation for their homes; Occidental Petroleum agreed to pay $129 million in compensations; and, in 1980, Congress passed the Comprehensive Environmental Response, Compensation and Liability Act, also known as the Superfund Act (see chapter 15).[4] It all started with someone paying attention, and it gathered momentum when someone else paid attention. Heroes are often forged in difficult, unsupported circumstances.

Another distressing turn of events is the coronavirus (COVID-19) pandemic. In addition to the health impact and economic shock, the world has gained experience with lower-carbon ways of being. According to Bob Leonard, coauthor of *Finite Earth Economy*, the admonition to "stay home, stay safe" was a prescription to significantly reduce fossil fuel consumption. Technology laggards, myself included, gained experience with teleconferencing. Telemedicine and online education both moved up the learning curve. More companies and employees gained work-at-home experience. Politicians have shown an occasional ability to cooperate, and scientists have gained renewed credibility.[5]

Could COVID-19 be an accelerating factor on the road to dealing with our other environmental challenges, especially Climate Change? Yes! Oil embargoes in the 1970s by the Organization of Petroleum Exporting Countries caused people with cars to wait in lines around the block for gasoline in the short run. In the long run, the crisis gave me—and thousands of other professionals—careers in energy efficiency. COVID-19 can profoundly shift our collective ways. Of course, that is at least partly up to you and me. Think Handprint opportunities.

Sometimes the challenge to create real constructive change seems daunting, overwhelming, or just plain hopeless. *We do the best we can, honoring our values, dedicated to the journey as much as the outcome.* Eventually something happens: A teacher. A new idea. An invitation. A distraction. An alignment of experience and opportunity. A maturing technology. An election, or two, or a thousand. Perhaps a young girl decides to sit on the steps of the Swedish parliament. Maybe one person joins her. Subtly—or suddenly—something changes. We make a difference.

HANDPRINT OPPORTUNITIES

20.1 Plan to create Handprints. Look for opportunities in your life events—like birthdays, vacations, community events, weddings, and graduations.

20.2 Convene a Handprint workshop or party. Ask participants create their own past and future Handprints on paper. Choose occasions like Earth Day in April, Earth Overshoot Day, the fall equinox in September, or the winter solstice in December. Celebrate each person's accomplishments and intentions. Bring lots of planet-friendly food—and reusable cutlery and dishes.

20.3 Respond to disasters with environmental intention.
- If the wind damages the roof, add insulation.
- If the water heater springs a leak, get a demand (tankless) water heater.
- If your work is miserable, look for environmentally friendly employment.
- If your personal economy goes south, move to a smaller house or an apartment.
- If the landlord evicts you, share space with others.
- When the car quits, buy an electric car—or go without.
- When disaster strikes your community, rebuild it green.

IN YOUR JOURNAL

How will *you* become a hero in your own story?

21

WE CAN DO THIS!

Perhaps you know someone like Milt Markewitz. After leaving IBM, he studied whole systems design at Antioch University. He is an energetic but soft-spoken man in his seventies who lobbies educators to teach in the way of living systems. He helps organize World Parliament of Religion events that include indigenous people and their perspectives. Perhaps your life has, or will have, a similar arc.

Whenever Milt speaks to a group, he shows a timeline graphic from his book *Language of Life*.[1] A line, starting high to the left, suggests a stable relationship with the earth's resources from prehistory until the agricultural and industrial ages. Then it dives downward exponentially, unsustainably. Population, resource extraction, waste, and pollution are driving us toward species extinction and global warming.

That line could continue downward, *or* . . .

A dotted projection from now toward the future shows us transforming that disastrous trajectory into an upward "Curve of Hope." It leads to a time of "ecological and social harmony." However, we must think and do things differently in ways that create "exponential improvements."

How can we make that dotted line in the Curve of Hope real?

TWELVE FUTURE COLLECTIVE HANDPRINTS

What is possible, even reasonable, to strive and hope for looking forward? Even when breaking news is discouraging, we have a right to be proud of the many Handprints we have already created. National parks. Efficient buildings. Cost-effective wind and solar energy. Electric cars. Recovered whales and turtles. Marine sanctuaries. Organic agriculture . . . These ac-

189

complishments yield *much* more than the sum of their parts. They form the foundation of a healthier world. I can say, with the confidence of one who has been eating and breathing sustainability for decades, that we have at least twelve future collective Handprints well within our grasp.

Within Our Grasp: A Green New Deal. Legislation is passed that promotes mutually reinforcing environmental, social, and economic objectives.

How we get there: We persist in developing earth-friendly and people-practical strategies and technologies, such as those described in these pages. We engage the legislative processes by supporting advocacy organizations and offering testimony—online, in print, in hearings, in person, and on the streets. We vote. We elect representatives who care. We run for office.

GREEN NEW DEAL GOALS

It is the duty of the Federal Government to create a Green New Deal—

    (A) to achieve net-zero greenhouse gas emissions through a fair and just transition for all communities and workers;

    (B) to create millions of good, high-wage jobs and ensure prosperity and economic security for all people of the United States;

    (C) to invest in the infrastructure and industry of the United States to sustainably meet the challenges of the 21st century;

    (D) to secure for all people of the United States for generations to come—

        (i) clean air and water;

        (ii) climate and community resiliency;

        (iii) healthy food;

        (iv) access to nature; and

        (v) a sustainable environment; and

    (E) to promote justice and equity by stopping current, preventing future, and repairing historic oppression of indigenous peoples, communities of color, migrant communities, deindustrialized communities, depopulated rural communities, the poor, low-income workers, women, the elderly, the unhoused, people with disabilities, and youth (referred to in this resolution as "frontline and vulnerable communities").

Source: "H. Res. 109—Recognizing the Duty of the Federal Government to Create a Green New Deal," Congress.gov, Library of Congress, accessed June 13, 2020, https://www.congress.gov/bill/116th-congress/house-resolution/109/text. Resolution introduced in Congress by Rep. Alexandria Ocasio-Cortez, February 7, 2019.

Within Our Grasp: Clean Rivers. The Mississippi, Los Angeles, and Ganges Rivers are healthy for fish and people.

How we get there: We inspire young and old with stories of the cleanups of the Hudson, Rhine, and Cuyahoga Rivers. Volunteers support river cleanups. The actions of water protectors amplify our concern for water. Trees are planted along the rivers. Financial and government support flow to organizations such as Living Lands and Waters (industrial scale cleanups), The Nature Conservancy (Mississippi River Basin), The Friends of the Los Angeles River, and Ganga (Ganges River) Action Parivar.

Within Our Grasp: Cleaner Oceans. The net flow of plastics into our oceans is reversed.

How we get there: We reduce the use of plastic packaging and buy more bulk food. We ban most onetime uses of the plastic bag and plastic straw. We develop technologies to cost-effectively recover plastic from rivers and ocean gyres.

Within Our Grasp: Renewable Energy Matures. Eighty percent of our power comes from renewable energy. Large coal and nuclear power plants and even natural gas generators are marginalized by the falling cost of wind and solar power.

How we get there: Individuals, universities, businesses, states, and countries fully commit to renewable energy. Activists continue to highlight the unsustainable costs of fossil and nuclear power. Investors shift their money from fossil and nuclear companies to renewable energy companies. Renewable researchers and engineers remain persistent. Politicians and bureaucrats implement policies favoring development of small and distributed energy resources. States and nations put a price on carbon.

Within Our Grasp: Evolved Buildings and Appliances. Smart appliances and distributed renewable energy supplies keep the power grid stable. Net zero energy buildings abound.

How we get there: Consumers buy Energy Star devices and homes and install LED lighting everywhere. Utilities and their customers cooperate to develop and use distributed energy resources. LEED buildings essentially set the sustainability standard for all new buildings. Cities offer incentives and otherwise promote efficiency for existing buildings.

Within Our Grasp: Transformed Mobility. Efficient electric cars, bicycles, and high-speed intercity rail are the norm. Internal combustion engines and gasoline stations are rare.

How we get there: People consider the environment in their choice of transportation. Transportation planners and companies continue to innovate. Cities discourage cars downtown. Millennials and Generation Z (those born in 1996 and later) exercise their collective economic and political will.

Within Our Grasp: Sustainable Business. Twenty percent of all business enterprises have the attributes of benefit corporations. The tipping point is reached when 10 percent of customers and stockholders vote sustainably with their dollars.

How we get there: Customers buy from resale businesses, benefit corporations (or B Corps), and other sustainable enterprises. Customers demand brands that represent planetary stewardship. Businesses internally use carbon offsets, Water Restoration Certificates, carbon pricing, and sustainability reporting.

Within Our Grasp: Population Stabilizes. The earth's human population growth levels off through peaceful and life-affirming processes. Al Gore, in *Our Choice* (2009), predicted that our population would stop growing midcentury at about 9.1 billion people. We can do better.

How we get there: Sustainable development ensures the security of billions. Girls are educated worldwide. Educated and empowered women choose the size of their families.

Within Our Grasp: Indigenous Respect. Corporations, governments, and individuals act in the best interests of indigenous people and the earth. Indigenous people around the world, and their teachings, are held in high regard. Indigenous people are stewards of many parks around the world.

How we get there: Environmentalists support indigenous people in confronting abuses of the land. Water protectors continue to resist environmentally insensitive development. Indigenous people sustainably manage land where they live. Consumers buy fair trade products and participate in boycott campaigns. Media report accordingly. The Earth Charter and the seventeen UN goals for sustainable development are referred to again and again in creating laws and treaties.

Within Our Grasp: Humans Regain Respect for Nature. Environmental literacy translates into environmentally sensitive democracy. Voters understand their environmental best interests.

How we get there: We read environmentally sensitive stories to children and make environmentally inspiring movies. We fund and attend outdoor schools. We insist that our schools require environmental literacy and embrace education for sustainable development. Young people choose lower-carbon lifestyles.

Within Our Grasp: Regenerative Agriculture. Farmers and ranchers enhance soil around the world using regenerative practices that sequester carbon from the atmosphere.

How we get there: Informed and empowered consumers buy Regenerative Organic Certified products. Passionate farmers, ranchers, and land stewards continue to develop regenerative strategies. Large companies and universities find ways to study, support, and profit from regenerative farming and ranching practices.

Within Our Grasp: Paris Works. "The [2015] Paris Agreement's central aim," according to the United Nations Climate Change website, "is to strengthen the global response to the threat of climate change by keeping a global temperature rise this century well below 2 degrees Celsius above pre-industrial levels and to pursue efforts to limit the temperature increase even further to 1.5 degrees Celsius. Additionally, the agreement aims to increase the ability of countries to deal with the impacts of climate change, and at making finance flows consistent with a low GHG [greenhouse gas] emissions and climate-resilient pathway."[2]

How we get there: Cities, subgovernment entities, and island nations ensure continued momentum in reducing carbon emissions. Europe maintains its role as the nexus of environmental leadership. Earth-friendly technologies and applications continue to evolve. Green investment flows to the developing world through nongovernmental organizations, carbon offsetting, and sustainable development. Environmental practices within China mature. United Nations Environmental Program diplomats persistently take stock every five years. Synergy: alignment of the 194 signers of the Paris Agreement creates a system more effective than the sum of its parts. The United States rejoins the agreement—because reversing global warming is in the national interest.

THE PROPHECY OF JAKE SWAMP

We certainly have grounds for concern with large ice sheets breaking off the Antarctic ice shelf, record-breaking temperatures, and heat-driven storms like Katrina, Sandy, Irma, Iota, Maria, and Michael. I do not know whether we will meet our Paris Agreement aspiration of holding our global temperature rise from the industrial revolution to 1.5 degrees Celsius. I cannot prescribe one medicine that will cure our global ailments. At some point, I will have to let go of my efforts to create sustainability. For those challenges, I keep the prophecy of Jake Swamp close to my heart.

Around 2005, I attended a workshop during the Village Building Convergence led by Jon Young. He shared a message from Jake Swamp, a respected elder who lived in the Mohawk Nation on the St. Lawrence River between the United States and Canada.[3] I will now pass that vision of the future on to you. Jacob "Tekaronhianeken" Swamp, who died in 2010, left us all with this prophecy of hope:

> There will be a time when we treat each other respectfully and our communities are strong and we live in harmony with the earth. That time will be two hundred years from now. The dysfunctional institutions we know now will not be taken down by force. They will fall away because no one uses them anymore.

This is what we are working to achieve, even though we may not see its fruition. You and I do not have to do everything needed to heal the planet; we just need to do our part.

HANDPRINT OPPORTUNITIES

21.1 Vote for representatives who are willing to help recover the land, reverse global warming, and reclaim the future.

21.2 Encourage others to vote. Voting increases our power to make a difference. It is an effective way to say we matter and our opinions matter. Check out Vote Forward to reach out to chronically underrepresented groups.

Appendix 1

DRAWDOWN SOLUTIONS VIS-À-VIS HANDPRINT OPPORTUNITIES

How do the "Handprint opportunities" in *Our Environmental Handprints* complement the "solutions" in *Drawdown* by Paul Hawken? Handprint opportunities address a broad range of environmental issues, while *Drawdown* solutions apply specifically to drivers of climate change. By sorting, sector by sector, we can see how the policy and engineering objectives of *Drawdown* solutions are supported by the individual actions suggested by Handprint opportunities. Put differently, both the solutions and opportunities below are all proactive responses to climate change. (The numbers refer to designations within each text.)[1]

BUILDINGS AND CITIES

Drawdown **Solutions.** 27 District Heating. 31 Insulation. 33 LED Lighting (Household). 42 Heat Pumps. 44 LED Lighting (Commercial). 45 Building Automation. 54 Walkable Cities. 57 Smart Thermostats. 58 Landfill Methane. 59 Bike Infrastructure. 61 Smart Glass. 71 Water Distribution. 73 Green Roofs. 79 Net Zero Buildings. 80 Retrofitting.

Handprint Opportunities. 12.1 Install LED lights. 12.2 Choose Energy Star appliances. 12.3 Use clothes drying racks and hanging bamboo rods. 12.4 Use a food box. 12.5 Buy used house parts. 12.6 Post "What's green about this room?" signs. 12.7 Eco-retrofit your home. 12.8 Move to a more earth-friendly home. 12.10 Live in a solar-friendly house. 12.11 Promote earth-friendly energy in your community. 19.2 Create climate neutral action plans.

ELECTRICITY GENERATION

***Drawdown* Solutions.** 2 Wind Turbines (Onshore) Electricity Generation. 8 Solar Farms. 10 Rooftop Solar. 18 Geothermal. 20 Nuclear. 22 Wind Turbines (Offshore). 25 Concentrated Solar. 29 Wave and Tidal. 30 Methane Digesters (Large). 34 Biomass. 41 Solar Water. 48 In-Stream Hydro. 50 Cogeneration. 64 Methane Digesters (Small). 68 Waste-to-Energy. 76 Micro Wind. 77 Energy Storage (Distributed). 77 Energy Storage (Utilities). 77 Grid Flexibility. 78 Microgrids.

Handprint Opportunities. 5.4 Invest in Clean 200 companies. 5.5 Divest from Carbon Underground 200 companies. 14.1 Buy renewable power. 14.2 Support your state citizens' utility board. 14.3 Work for a wind, solar, or efficiency company. 14.4 Promote energy efficiency. 14.5 Donate to market transformation organizations. 14.6 Share the land you love with windmills. 14.7 Buy and use a solar cooker. 14.8 Install solar collectors. 14.9 Go "community solar." 14.10 Use solar-powered devices. 19.1 Advocate for 100 percent renewable energy. 19.2 Create climate neutral action plans. 19.3 Put a price on carbon. 19.6 Ensure that the United States fully recommits to the Paris Agreement. 19.8 Find employment in sustainable development or climate change mitigation.

FOOD

***Drawdown* Solutions.** 3 Reduced Waste. 4 Plant-Rich Diet. 9 Silvopasture. 11 Regenerative Agriculture. 14 Tropical Staple Trees. 16 Conservation Agriculture. 17 Tree Intercropping. 19 Managed Grazing. 21 Clean Cookstoves. 23 Farmland Restoration. 24 Improved Rice Cultivation. 28 Multistrata Agroforestry. 53 System of Rice Intensification. 60 Composting. 65 Nutrient Management. 67 Farmland Irrigation. 72 Biochar.

Handprint Opportunities. 4.4 Share your garden and orchard surplus. 10.1 Support The Land Institute. 10.2 Buy agricultural carbon offsets. 10.3 Spread compost on bare or compacted land. 10.4 Study botany and indigenous food cultivation. 10.5 Use sustainable farming practices. 10.6 Become certified as an organic grower. 10.7 Practice "mob grazing." 11.1 Grow food in a garden. 11.2 Buy food at farmers markets. 11.3 Become a community-supported agriculture member. 11.4 Go gleaning. 11.6 Buy Demeter Certified Biodynamic products. 11.7 Serve vegetables, fruit, and nuts. 11.8 Serve organic food. 11.9 Serve perennial food. 11.10 Serve raw food and fermented food. 11.11 Explore plant-based alternatives to animal products. 11.12 Create earth-friendly recipes. 11.13 Let go of beef. 11.14 Teach others to serve earth-friendly food. 11.15 Write online reviews.

LAND USE

Drawdown **Solutions.** 5 Tropical Forests. 12 Temperate Forests. 13 Peatlands. 15 Afforestation. 35 Bamboo. 38 Forest Protection. 39 Indigenous Peoples' Land Management. 51 Perennial Biomass. 52 Coastal Wetlands.

Handprint Opportunities. 1.1 Introduce children to trees. 1.6 Plant trees. 1.7 Donate to plant trees. 1.8 Volunteer for your local arboretum or botanical garden. 1.9 Endow the Tree Fund. 1.10 Research how trees sequester carbon. 1.12 Offset your carbon footprint. 1.13 Develop a carbon offset policy at work. 2.1 Buy plant-based soap. 2.2 Buy earth-friendly toilet paper. 2.3 Give cloth napkins and shopping bags as gifts. 2.4 Carry and use reusable spoon, fork, or spork and bowl. 2.5 Buy bamboo, machine-washable merino wool, ramie, organic hemp, or organic cotton clothing. 2.6 Buy classy used clothing. 15.1 Visit a national park. 15.2 Support trail building and maintenance. 15.3 Support indigenous people who protect the land. 15.4 Donate to African Parks. 15.5 Bequest your land to a conservation organization. 15.6 Form a land trust. 15.7 Support organizations that protect endangered species. 15.8 Create a certified wildlife habitat. 15.9 Use permaculture principles. 15.10 Use mycelium (mushrooms) to treat biohazards. 15.11 Help expand the fungi knowledge base. 15.12 Promote brownfield recovery. 15.13 Adopt a park.

MATERIALS

Drawdown **Solutions.** 1 Refrigerant Management. 36 Alternative Cement. 46 Water Saving—Home. 47 Bioplastic. 55 Household Recycling. 56 Industrial Recycling. 70 Recycled Paper.

Handprint Opportunities. 1.2 Buy 100 percent recycled printer paper. 1.3 Move to online bill paying. 1.4 Switch to double-sided (duplex) printing. 1.5 Switch to reading e-books. 2.2 Buy treeless toilet paper. 4.3 Donate reusables to a reselling outlet. 4.5 Use Freecycle or Craigslist. 12.8 Salvage tear-down buildings. 17.4 Propose to ban plastic bags. 18.1 Buy and sell clothes and furniture from a resale store. 18.2 Rent instead of buying tools and cars. 18.3 Buy goods made from recycled materials. 18.4 Recycle paper, metal, glass, and acceptable plastics through recycling programs. 18.5 Use TerraCycle Zero Waste Boxes. 18.7 Buy products with Cradle to Cradle Certified materials.

TRANSPORTATION

Drawdown **Solutions.** 26 Electric Vehicles. 32 Ships. 37 Mass Transit. 40 Trucks. 43 Airplanes. 49 Cars. 63 Telepresence. 66 High-speed Rail. 69 Electric Bikes. 74 Trains. 75 Ridesharing.

    Handprint Opportunities. 13.1 Walk or ride a bike. 13.2 Arrange online meetings. 13.3 Carpool. 13.4 Trade in your gasoline car for an electric car. 13.5 Live without owning a car. 13.6 Take children on the bus. 13.7 Drive a bus or a train. 13.8 Participate in public transportation planning.

WOMEN AND GIRLS

Drawdown **Solutions.** 6 Educating Girls. 7 Family Planning. 62 Women Smallholders.

    Handprint Opportunities. 7.7 Support organizations that educate girls and women. 7.8 Support organizations that support family planning.

Appendix 2

UNITED NATIONS SUSTAINABLE DEVELOPMENT GOALS

Sustainable Development Goal (SDG)	Example SDG Target Highlight(s) (paraphrased to emphasize environmental actions that individuals can support)	Applicable Chapters in Our Environmental Handprints
1. Poverty	1.4 Provide access to microfinance. 1.5 Reduce poor people's vulnerability to extreme climate-related events and other shocks and disasters.	Chapter 6. Work for Environmental Justice Chapter 7. Support Sustainable Development
2. Hunger	2.3 Double the productivity and income of small-scale food producers. 2.5 Maintain the genetic diversity of seeds.	Chapter 10. Honor the Soil Chapter 11. Partner with Earth-Friendly Farmers
3. Health	3.9 Reduce death and illness due to hazardous chemicals and soil pollution.	Chapter 15. Protect and Recover the Land
4. Education	4.5 Eliminate gender disparity in education. 4.6 Ensure access to literacy and numeracy education for all youth. 4.7 Teach lifestyle and nonviolence skills needed to promote sustainable development.	Chapter 2. Little Things Add Up Chapter 7. Support Sustainable Development Chapter 8. Teach Ecology Chapter 9. Nurture a Lifelong Relationship with Nature
5. Gender equality	5.1 End all forms of discrimination against all women and girls everywhere. 5.6 Provide universal access to sexual and reproductive health services.	Chapter 7. Support Sustainable Development Chapter 13. Move with Graceful Economy.
6. Water and sanitation	6.a Invest in water efficiency, water treatment, and desalinization technologies.	Chapter 16. Revive Our Rivers
7. Energy	7.2 Increase the share of renewable energy in the global energy mix. 7.3 Double global energy efficiency.	Chapter 14. Insist on Earth-Friendly Energy Chapter 19. The Paris Agreement
8. Economy	8.8 Ensure safe and secure working environments.	Chapter 6. Work for Environmental Justice
9. Infrastructure	9.4 Upgrade infrastructure and retrofit industries to increase resource use efficiency.	Chapter 12. Live and Work in Modest Elegance Chapter 13. Move with Graceful Economy
10. Inequity	10.7 Facilitate orderly, safe, planned, and well-managed immigration policies.	Chapter 6. Work for Environmental Justice

	Target	Chapters
11. Cities	11.6 Reduce per capita environmental impact of cities.	Chapter 12. Live and Work in Modest Elegance Chapter 13. Move with Graceful Economy Chapter 19. The Paris Agreement
12. Consumption and production	12.5 Reduce waste generation through prevention, reduction, recycling, and reuse.	Chapter 5. Invest in Sustainability Chapter 11. Partner with Earth-Friendly Farmers
	12.c Remove market distortions resulting from fossil fuel subsidies.	Chapter 18. A Circular Economy
13. Climate change	13.1 Support actions taken under the UN Framework Convention on Climate Change adopted after the 1992 Rio de Janeiro Earth Summit.	Chapter 1. Handprints and Footprints Chapter 5. Invest in Sustainability Chapter 7. Support Sustainable Development
	13.2 Integrate climate change measures into national policies, strategies, and planning.	Chapter 10. Honor the Soil Chapter 11. Partner with Earth-Friendly Farmers
	13.3 Improve education, awareness raising, and human and institutional capacity on climate change mitigation, adaptation, impact reduction, and early warning.	Chapter 12. Live and Work in Modest Elegance Chapter 13. Move with Graceful Economy Chapter 14. Insist on Earth-Friendly Energy
	13.4 Mobilize $100 billion annually from all sources in developed countries to address the needs of developing countries.	Chapter 19. The Paris Agreement
14. Rivers, lakes, and oceans	14.1 Prevent and significantly reduce marine pollution of all kinds, in particular from land-based activities.	Chapter 2. Little Things Add Up Chapter 16. Revive Our Rivers
	14.5 Conserve at least 10 percent of coastal and marine areas.	Chapter 17. Conserve Our Oceans Chapter 18. A Circular Economy
15. Land	15.1 Ensure the conservation, restoration, and sustainable use of terrestrial and inland freshwater ecosystems and their services, in particular forests, wetlands, mountains, and drylands, in line with international agreements.	Chapter 15. Protect and Recover the Land Chapter 16. Revive Our Rivers
	15.7 End trafficking of protected species.	
16. Peace and justice	16.7 Ensure responsive, inclusive, participatory, and representative decision making at all levels.	Chapter 3. Nurture Community
17. Partnership	17.7 Diffuse environmentally sound technologies.	Chapter 6. Work for Environmental Justice Chapter 19. Paris

NOTES

INTRODUCTION

1. Greta Thunberg, "Transcript: Greta Thunberg's Speech at the UN Climate Action Summit," National Public Radio, September 23, 2019, https://www.npr.org/2019/09/23/763452863/transcript-greta-thunbergs-speech-at-the-u-n-climate-action-summit.

CHAPTER 1

1. *Merriam-Webster*, s.v. "handprint," accessed December 29, 2019, https://www.merriam-webster.com/dictionary/handprint.

2. "The Handprint Idea," Centre for Environmental Education, Ahmedabad, India, accessed December 26, 2019, https://www.handprint.in/the_handprint_idea.

3. Greg Norris, "Introducing Handprints: A Net-Positive Approach to Sustainability," Harvard Extension School, accessed December 21, 2019, https://www.extension.harvard.edu/introducing-handprints.

4. Rocky Rohwedder, *Ecological Handprints* (self-pub., 2016), https://ecologicalhandprints.atavist.com/ecological-handprints?promo.

5. Jon Biemer, Willow Dixon, and Natalia Blackburn, "Our Environmental Handprint: The Good We Do," Institute of Electrical and Electronic Engineers, 2013 IEEE Conference on Technologies for Sustainability, Portland, Oregon.

6. Erv Evans, "Tree Facts," University of North Carolina Cooperative Extension, accessed July, 20, 2019.

7. Steve Curwood, "Nobel Prize Winner Wangari Maathai Remembered," *Living on Earth*, August 10, 2012, http://www.loe.org/blog/blogs.html/?seriesID=1&blogID=20.

8. Alex Steffen, ed., *WorldChanging: A User's Guide for the 21st Century* (New York: Abrams, 2011), 346.

9. "Structure," Plant for the Planet, accessed July 24, 2015, http://www.plant-for-the-planet.org/en/about-us/structure.

10. Mark Tutton, "The Most Effective Way to Tackle Climate Change? Plant 1 Trillion Trees," CNN, updated April 17, 2019, https://www.cnn.com/2019/04/17/world/trillion-trees-climate-change-intl-scn/index.html.

11. Jonathan Amos, "Earth's Trees Number 'Three Trillion,'" *BBC News*, September 3, 2015, http://www.bbc.com/news/science-environment-34134366.

12. John Vidal, "This German Teen Is Leading a Global Plan to Plant a Trillion Trees," *Huffpost*, March 27, 2018, https://www.huffpost.com/entry/tree-planting-felix-finkbeiner_n_5ab3b850e4b0decad0478b8c.

13. Douglas Brinkley, *Rightful Heritage: Franklin D. Roosevelt and the Land of America* (New York: HarperCollins, 2016), 472.

14. Mathis Wackernagel and William Rees, *Our Ecological Footprint: Reducing Human Impact on the Earth* (Gabriola Island, BC: New Society, 1996).

15. "Sustainable Development," Global Footprint Network, accessed July 19, 2019, https://www.footprintnetwork.org/our-work/sustainable-development.

16. "Our Past and Our Future," Global Footprint Network, accessed November 10, 2020, https://www.footprintnetwork.org/about-us/our-history.

17. Tali Sharot, *The Influential Mind* (New York: Henry Holt, 2017), 66.

18. "Greenhouse Gas Emissions from a Typical Passenger Vehicle," Environmental Protection Agency, EPA-420-F-14-040a, May 2014, https://nepis.epa.gov/Exe/ZyPDF.cgi?Dockey=P100LQ99.pdf.

19. "State Plastic and Paper Bag Legislation," National Conference of State Legislators, September 29, 2020, https://www.ncsl.org/research/environment-and-natural-resources/plastic-bag-legislation.aspx.

20. Mathis Wackernagel (coauthor of *Ecological Footprint: Managing the Biocapacity Budget*), email message to author, November 8, 2020.

21. "U.S. Farmers Earn World's First Carbon Credits for Rice Cultivation," American Carbon Registry, June 14, 2017, https://americancarbonregistry.org/news-events/news/u-s-farmers-earn-world2019s-first-carbon-credits-from-rice-cultivation-conservation-practices-result-in-credible-sustainability-benefits-including-reduced-greenhouse-gas-emissions-and-savings-in-water-and-energy-use.

22. "Ghana Clean Water Project," Native Energy, accessed June 1, 2019, http://www.nativeenergy.com/ghana-clean-water-project.html.

23. "Calculate Your Footprint and Reduce Your Impact," Conservation International, accessed July 23, 2019, https://www.conservation.org/carbon-footprint-calculator#.

24. "Carbon Emissions Reduction Project in the Forest Corridor Ambositra-Vondrozo Forest Corridor (COFAV)—Madagascar: Climate, Community and Biodiversity Standards," Conservation International, January 2014, p. 4, https://s3.amazonaws.com/CCBA/Projects/Reduced_Emissions_from_Deforestation_in_

the_Ambositra-Vondrozo_Forest_Corridor_(COFAV)-Madagascar_Project/Vali dation/Update/COFAV+CCBS+PD+2014_01_07_validated.pdf.

25. Mohammad Asrafur Rahman, "Quantifying the Cooling Effectiveness of Urban Street Trees in Relation to Their Growth," Tree Fund, January 28, 2016, http://www.treefund.org/archives/11571.

26. "What Is Your Ecological Footprint?" Global Footprint Network, accessed November 11, 2020, https://www.footprintcalculator.org.

27. "Individuals and Families," Terrapass, accessed January 5, 2020, https://www.terrapass.com/product-category/individuals.

CHAPTER 2

1. "EWG's Guide to Healthy Cleaning," Environmental Working Group, accessed August 19, 2019, https://www.ewg.org.

2. "Handcrafted Soap and Cosmetic Guild," Handcrafted Soap and Cosmetic Guild, accessed August 19, 2019, https://www.soapguild.org.

3. Brittney Morgan, "We Tried Tree-Free Toilet Paper, and Here's What We Thought," *Apartment Therapy*, October 5, 2017, https://www.apartmenttherapy.com/rebel-green-tree-free-toilet-paper-review-250020.

4. "Who Gives a Crap," Who Gives a Crap, accessed October 17, 2019, https://us.whogivesacrap.org.

5. "Bamboo Clothing Gets 5 Stars," Cool-Organic-Clothing.com, accessed January 4, 2020, https://www.cool-organic-clothing.com/bamboo-clothing.html.

6. "Get the Scoop on Organic Hemp," Cool-Organic-Clothing.com, accessed January 4, 2020, https://www.cool-organic-clothing.com/organic-hemp.html; "Why Is Hemp Illegal?" Hemp Helps, updated May 2, 2019, https://www.hemphelps.org/why-hemp-is-illegal.

7. "Ramie Fiber—From Silky Soft to Coarse Denim," Cool-Organic-Clothing.com, accessed January 4, 2020, https://www.cool-organic-clothing.com/ramie-fiber.html.

8. "Organic Cotton Is Better for the Environment," Textile Exchange, accessed August 19, 2019, http://aboutorganiccotton.org/environmental-benefits.

9. "Organic by Choice," Textile Exchange, accessed August 19, 2019, http://aboutorganiccotton.org.

10. Marjorie van Elven, "How Sustainable Is Recycled Polyester?" Fashion-United, November 15, 2018, https://fashionunited.uk/news/fashion/how-sustainable-is-recycled-polyester/2018111540000.

11. "Patagonia, Inc.," B Lab, accessed April 2, 2016, https://www.bcorporation.net/community/patagonia-inc.

12. "ReCrafted, These Are Clothes Made from Other Clothes," Patagonia, accessed December 30, 2019, https://wornwear.patagonia.com/shop/recrafted.

13. "Thankful Rose," Etsy, accessed December 30, 2019, https://www.etsy.com/shop/ThankfulRose.

14. "Popular Items for Upcycled Clothing," Etsy, accessed December 30, 2019, https://www.etsy.com/market/upcycled_clothing.

15. "We Are What We Wear," ThredUp, accessed December 30, 2019, https://www.thredup.com/p/about.

16. Home page, Handcrafted Soap and Cosmetic Guild, accessed August 19, 2019, https://www.soapguild.org.

CHAPTER 3

1. William Hageman, "Tool Libraries Lend DIYers a Hand," *Chicago Tribune*, July 19, 2010, http://articles.chicagotribune.com/2010-07-19/classified/sc-home-0719-tool-library-20100719_1_lending-library-tool-libraries-general-home-repair.

2. Home page, Toe River Skill Exchange, accessed January 30, 2020, https://hourworld.org/bank/?hw=1551.

3. Home page, Time Banks, accessed January 31, 2020, https://timebanks.org/.

4. "Directory of Time Banks," Timebanks USA, accessed January 31, 2020, http://community.timebanks.org/.

5. Katie Gilbert, "Communes Still Thrive Decades After the 60's, but Economy Is a Bummer, Man," *Aljazeera America*, December 7, 2014, http://america.aljazeera.com/multimedia/2014/12/communes-still-thrivedecadesafterthe60sbuteconomy isabummerman.html. "Fellowship for Intentional Community," Fellowship for Intentional Community, accessed December 31, 2016, http://www.ic.org/.

6. "About Dancing Rabbit Ecovillage," Dancing Rabbit Ecovillage, accessed January 3, 2017, http://www.dancingrabbit.org/about-dancing-rabbit-ecovillage/; "Come Visit and Experience Sustainable Community Living First Hand," Dancing Rabbit Ecovillage, accessed December 31, 2016, http://www.dancingrabbit.org/visit/.

7. "Twenty Years of Lessons Learned," Ecovillage Ithaca, accessed January 1, 2017, http://ecovillageithaca.org/download/evi-20-years-of-lessons-learned/.

8. "Live," Ecovillage Ithaca, accessed January 2, 2017, http://ecovillageithaca.org/live/.

9. "Enright Ridge Urban Ecovillage," Fellowship for Intentional Community, accessed January 4, 2017, http://www.ic.org/directory/enright-ridge-urban-eco village/.

10. "Kailash Ecovillage," Kailash Ecovillage, accessed December 31, 2016, http://www.kailashecovillage.org/.

11. City Repair, *Placemaking Guidebook*, 2nd ed. (Portland, OR: City Repair Project, 2006), 14–15.

12. Project for Public Spaces, accessed October 10, 2020, https://www.pps.org/.

13. "About," Community Environmental Legal Defense Fund, accessed August 23, 2018, https://celdf.org/about/.

14. "Chapter 162 Protection of Natural Resources" (ordinance), Town of Wales, New York, adopted June 14, 2011, https://ecode360.com/28771798.

15. "Shapleigh, Maine, Residents Vote in Rights-Based Ordinance to Protect Their Water," (video) Community Environmental Legal Defense Fund, accessed October 20, 2020, https://celdf.org/2009/02/shapleigh-maine-residents-vote-in-rights-based-ordinance-to-protect-their-water/.

16. Associated Press, "Federal Judge Refuses to Block Jackson County's GMO Ban," *The Oregonian/OregonLive*, May 20, 2015, http://www.oregonlive.com/pacific-northwest-news/index.ssf/2015/05/federal_judge_refuses_to_block.html.

17. Luis Acosta, "Restrictions on Genetically Modified Organisms: United States," Law Library of the Library of Congress, March 2014, http://www.loc.gov/law/help/restrictions-on-gmos/usa.php.

18. "Ordinance 2013-1," County of Mora, State of New Mexico, April 29, 2013, https://therightsofnature.org/wp-content/uploads/pdfs/Mora_County_Ordinance_2013.pdf.

19. Gail Darrell, "Sangerville, Maine Adopts Community Bill of Rights Ordinance to Reject Transportation and Distribution Corridors," *The Dirt*, October 8, 2013, https://readthedirt.org/sangerville-maine-adopts-community-bill-of-rights-ordinance-to-reject-transportation-and-distribution-corridors.

20. "State and National Networks," Community Environmental Legal Defense Fund, accessed April 18, 2020, https://celdf.org/join-the-movement/where-we-work/state-national-networks/.

21. "Community Rights Do-It-Yourself Guide to Lawmaking," Community Environmental Legal Defense Fund, accessed April 18, 2020, http://celdf.org/wp-content/uploads/2019/11/DIY-Guide-2019-FINAL-2.pdf.

22. "Experience Chamainus Vancouver Island," Municipality of Cowichan, accessed December 27, 2016, http://www.chemainus.com/.

23. "Visit a Repair Café," Repair Café, accessed October 20, 2019, https://repaircafe.org/en/visit/.

24. Ted Sickinger, "Opponents of Water Bottling in Cascade Locks Say Nestlé Hid Campaign Contributions," *The Oregonian/OregonLive*, May 14, 2016, http://www.oregonlive.com/pacific-northwest-news/index.ssf/2016/05/opponents_of_water_bottling_pl.html; Chris Holmstrom and Hanna Button, "Voters Pass Measure to Keep Nestlé Out of Cascade Locks," *KOIN 6*, LIN Television Corporation, May 17, 2016, https://www.koin.com/news/voters-pass-measure-to-keep-nestle-out-of-cascade-locks/.

CHAPTER 4

1. "History—More Than 40 Years of Ministry," Society of St. Andrew, accessed August 28, 2019, https://endhunger.org/history/.

2. "Society of St. Andrew," Society of St. Andrew, accessed August 28, 2019, https://endhunger.org/.

3. "About Us," Urban Gleaners, accessed July 21, 2014, http://urbangleaners. org/about-us.

4. "Feeding America, 2018 Annual Report," Feeding America, accessed October 30, 2019, https://www.feedingamerica.org/sites/default/files/2018-12/2018%20 Feeding%20America%20Annual%20Report_0.pdf.

5. "2019 Sustainability Progress," Trader Joe's, January 1, 2020, https://www. traderjoes.com/announcement/2019-sustainability-progress.

6. "YouCaring Is Now Part of GoFundMe," GoFundMe, accessed October 21, 2020, https://www.gofundme.com/c/youcaring.

7. "Our Mission Is to Bring Creative Projects to Life," Kickstarter, accessed October 30, 2019, https://www.kickstarter.com/about?ref=global-footer.

8. Devin Thorpe, "Q&A with Indiegogo's Slava Rubin about Helping Entrepreneurs Raise $1.5 Billion," *Forbes*, September 13, 2018, https://www.forbes. com/sites/devinthorpe/2018/09/13/after-helping-people-raise-over-1-5b-a-con versation-with-indiegogos-slava-rubin/#1569177135fe.

9. "Solar Roadways," Indiegogo, accessed July 29, 2019, https://www.indi egogo.com/projects/solar-roadways#/.

10. Home page, Freecycle Network, accessed December 8, 2019, https://www. freecycle.org/.

11. Home page, Chip Drop, accessed July 7, 2014, https://getchipdrop.com/.

12. "Portland, OR," Craigslist, accessed October 31, 2019, https://portland. craigslist.org/.

13. "What Is BookCrossing?" BookCrossing, accessed October 31, 2019, https://www.bookcrossing.com/.

14. "Directory of Charities and Nonprofit Organizations," GuideStar, accessed October 21, 2020, https://www.guidestar.org/NonprofitDirectory.aspx?cat=3.

15. "2010 Annual Report," Friends of the Earth, October 21, 2020, https:// foe.org/resources/?s=&publication-types%5B%5D=annual-reports&order=DESC.

16. "Friends of the Earth International Financial Statement 2019," Friends of the Earth International, October 21, 2020, https://www.foei.org/wp-content/ uploads/2020/07/Financial-statements-2019-English.pdf.

17. "2017 Annual Report," Greenpeace, accessed October 14, 2020, https:// www.greenpeace.org/usa/2017-annual-report/.

18. "Origin Story," Pachamama Alliance, accessed June 25, 2020, https://www. pachamama.org/about/origin.

19. "George B. Dorr," National Park Service, accessed October 21, 2020, https://www.nps.gov/people/george-b-dorr.htm. "Creation of Grand Teton National Park (A Thumbnail Sketch)," National Park Service, accessed October 21, 2020, http://www.nps.gov/grte/planyourvisit/upload/creation.pdf.

20. Stephanie Harris, "Greg Carr's Big Gamble," *Smithsonian*, May 2007, http://www.smithsonianmag.com/people-places/greg-carrs-big-gamble-153081070/?no-ist=&page=1.

21. "The Global Alliance for Wild Cats," Panthera International, accessed June 14, 2016, https://www.panthera.org/global-alliance.

22. Matthew Shaer, "Into Thin Air," *Smithsonian*, March 2016, https://www.smithsonianmag.com/issue/march-2016/.

23. "Stanford Launches $100 Million Initiative to Tackle Energy Issues," *Stanford Report*, January 12, 2009, https://news.stanford.edu/news/2009/january14/pie-011409.html; "Seed Grant Spotlight: Waste Not, Want Not," TomKat Center for Sustainable Energy, March 2, 2017, https://tomkat.stanford.edu/seed-grant-spotlight-waste-not-want-not; "Speedy Bikes," *Stanford Report*, March 2016, https://tomkat.stanford.edu/speedy-cycles; "XStream Spotlight," *Stanford Report*, September 2016, https://tomkat.stanford.edu/xstream-spotlight.

24. Naomi Klein, *This Changes Everything* (New York: Simon & Schuster, 2014), 230–55.

25. "Find Your Local Food Bank," Feeding America, accessed August 11, 2018, http://www.feedingamerica.org/find-your-local-foodbank/.

CHAPTER 5

1. "2014 Annual Report," US SIF and US SIF Foundation, accessed October 14, 2020, http://www.ussif.org/files/Publications/Annual_%20Report_14_FINAL.PDF. "2018 Report on Sustainable, Responsible and Impact Investing Trends," US SIF, accessed October 31, 2019, https://www.ussif.org/trends.

2. Zoe van Schyndel, "Go Green with Socially Responsible Investing," *Investopedia*, updated June 25, 2019, http://www.investopedia.com/articles/07/clean_and_green.asp.

3. Melissa D. Berry, "History of Socially Responsible Investing in the U.S.," Thomson Reuters, August 9, 2013, http://sustainability.thomsonreuters.com/2013/08/09/history-of-socially-responsible-investing-in-the-u-s/.

4. Shira Shoenberg, "Pension Politics: The History of Divestment in Massachusetts," MassLive, updated March 24, 2019, https://www.masslive.com/politics/2014/05/the_history_of_divestment_in_m.html.

5. "'Clean Trillion' Investment Goal in Sight and Achievable, New . . . ," Ceres, May 10, 2018, https://www.ceres.org/news-center/press-releases/clean-trillion-investment-goal-sight-and-achievable-new-ceres-report.

6. Bill McKibben, "Global Warming's Terrifying New Math," *Rolling Stone*, July 19, 2012, http://www.rollingstone.com/politics/news/global-warmings-terrifying-new-math-20120719.

7. "The Carbon Underground™—2017 Edition," Fossil Free Indices, accessed October 14, 2020, http://fossilfreeindexes.com/research/the-carbon-underground/.

8. "1000+ Divestment Commitments," Fossil Free (a project of 350.org), accessed August 29, 2019, https://gofossilfree.org/divestment/commitments/.

9. Reuters, "India Says Paris Deal Won't Affect Plans to Double Coal Output," *Guardian*, December 14, 2015, http://www.theguardian.com/environment/2015/dec/14/india-says-paris-climate-deal-wont-affect-plans-to-double-coal-output.

10. "Clean 200™," As You Sow, accessed February 4, 2020, https://www.asyousow.org/clean200.

11. "Carbon Clean 200™: Investing in a Clean Energy Future, 2019 Q1 Performance Update," As You Sow, February 2019, https://www.asyousow.org/report/clean200-2019-q1; "Sustainability Is in Our DNA," Prologis, accessed February 4, 2020, https://www.prologis.com/about/sustainable-industrial-real-estate.

12. "Investor Network on Climate Risk," Ceres, accessed July 12, 2015, http://www.ceres.org/investor-network/incr.

13. "About ICCR," Interfaith Center on Corporate Responsibility, accessed July 15, 2015, http://www.iccr.org/about-iccr.

14. "History of ICCR," Interfaith Center for Corporate Responsibility, accessed August 29, 2019, https://www.iccr.org/about-iccr/history-iccr.

15. Leslie Guevarra, "How Shareholder Activism Moved the Needle on Sustainability in 2011," *GreenBiz*, July 8, 2011, http://www.greenbiz.com/news/2011/07/08/how-shareholder-activism-moved-needle-sustainability-2011

16. Cyrus Nemati, "Shareholders Are Plugging Methane Leaks Themselves," As You Sow, June 1, 2018, https://www.asyousow.org/blog/2018/6/1/shareholders-are-plugging-methane-leaks-themselves.

17. Jeffery Jones, "StatOil Halts Multibillion-Dollar Oil Sands Project," *Globe and Mail*, October 25, 2014, http://www.theglobeandmail.com/report-on-busi ness/industry-news/energy-and-resources/statoil-halts-multibillion-dollar-alberta-project/article20790038/; Tim Brennan et al., "Addressing Climate Change through UUA Investments," (presentation), Unitarian Universalist General Assembly, Portland, Oregon, June 25, 2015.

18. "FirstEnergy Corporation: Request for Report on Carbon Asset Risk," As You Sow, November 30, 2017, https://www.asyousow.org/resolu tions/2017/11/30/firstenergy-corporation-request-for-report-on-carbon-asset-risk; Aaron Larson, "FirstEnergy Throws in the Towel on Coal Plants," *Power Magazine*, August 30, 2018, https://www.powermag.com/firstenergy-throws-in-the-towel-on-coal-plants/.

19. "Manufacturer Takeback Programs in the US," Electronics Takeback Coalition, updated March 2016, http://www.electronicstakeback.com/how-to-recycle-electronics/manufacturer-takeback-programs/; "Ocean Plastics," As You Sow, accessed May 7, 2020, https://www.asyousow.org/our-work/waste/ocean-plastics.

20. Larry Fink, "A Fundamental Reshaping of Finance," (letter to CEOs), BlackRock, January 2020, https://www.blackrock.com/us/individual/larry-fink-ceo-letter.

21. Greg McFarlane, "How BlackRock Makes Money," *Investopedia*, updated September 24, 2019, https://www.investopedia.com/articles/markets/012616/how-blackrock-makes-money.asp.

22. "Online Directory," The Forum for Sustainable and Responsible Investment, accessed February 8, 2018, http://www.ussif.org/AF_MemberDirectory.asp.

CHAPTER 6

1. "About US," Earth Charter International, accessed October 22, 2020, https://earthcharter.org/about-us/.

2. "The Earth Charter/Rio de Janeiro Declaration and the Oneness of Humanity," Baha'i International Community, March 4, 1992, https://www.bic.org/statements/earth-charterrio-de-janeiro-declaration-and-oneness-humanity.

3. "History," Earth Charter International, accessed October 22, 2020, https://earthcharter.org/about-us/history/.

4. "The Earth Charter," Earth Charter Commission, accessed May 6, 2020, https://earthcharter.org/wp-content/uploads/2020/03/echarter_english.pdf?x28510.

5. "Earth Charter in Action. Powering a Global Movement." Earth Charter International, accessed October 22, 2020, https://earthcharter.org/the-movement/.

6. *The Earth Charter*, Earth Charter International, approved June 29, 2000, https://earthcharter.org/wp-content/uploads/2020/03/echarter_english.pdf?x79755.

7. Dayna Jones, "Flint Water Crisis: The Importance of Building a Grassroots Environmental Justice Infrastructure," *Law at the Margins*, August 30, 2019, https://lawatthemargins.com/2553-2/.

8. "Flint, MI," Data USA, accessed November 6, 2019, https://datausa.io/profile/geo/flint-mi/.

9. Edwin Rios, "Meet 5 Everyday Heroes of Flint's Water Crisis," *Mother Jones*, January 27, 2016, https://www.motherjones.com/politics/2016/01/flint-water-crisis-lead-heroes/.

10. Staten Island Advance, "From the Archives: Trail of Broken Promises Litters History of Former Fresh Kills Landfill," updated March 4, 2019, https://www.silive.com/news/2013/09/from_the_archives_trail_of_bro.html. "Freshkills Park," New York City Parks, accessed October 22, 2020, https://www.nycgovparks.org/park-features/freshkills-park/about-the-site.

11. Martin Melosi, *Garbage in the Cities: Refuse, Reform, and the Environment* (Pittsburgh, PA: University of Pittsburgh Press, 2005).

12. "History of the RCRA," Environmental Protection Agency, accessed May 6, 2020, https://www.epa.gov/rcra/history-resource-conservation-and-recovery-act-rcra.

13. "National Overview: Facts and Figures on Materials, Wastes and Recycling," Environmental Protection Agency, accessed July 4, 2019, https://www.epa.gov/facts-and-figures-about-materials-waste-and-recycling/national-overview-facts-and-figures-materials.

14. Dennis Sumrak, "Navajos Will Never Forget the 1864 Scorched-Earth Campaign," *Wild West*, October 2012, reprinted in History Net, accessed November 7, 2019, https://www.historynet.com/navajos-will-never-forget-1864-scorched-earth-campaign.htm.

15. Gilbert King, "Where the Buffalo No Longer Roamed," *Smithsonian*, July 17, 2012, https://www.smithsonianmag.com/history/where-the-buffalo-no-longer-roamed-3067904/.

16. "1874 Expedition to the Black Hills," BH Visitor, accessed November 7, 2019, https://blackhillsvisitor.com/learn/1874-custer-expedition-to-the-black-hills/.

17. Robert Alvarez, "Uranium Mining and the U.S. Nuclear Weapons Program," Federation of American Scientists, November 14, 2013, https://fas.org/pir-pubs/uranium-mining-u-s-nuclear-weapons-program-3/.

18. Judith Graham, "Compensation at Last for Tribes that Lost Lands to Dams," *Chicago Tribune*, December 15, 2002, https://www.chicagotribune.com/news/ct-xpm-2002-12-15-0212150473-story.html.

19. Hannah Nguyen, "Church Rock: The Forgotten Nuclear Disaster," (submitted as coursework), Stanford University, March 12, 2019, http://large.stanford.edu/courses/2019/ph241/nguyen-h2/.

20. Dina Gilio-Whitaker, *As Long as the Grass Grows* (Boston: Beacon, 2019), 49–52.

21. Gilio-Whitaker, *As Long as the Grass Grows*, 106.

22. Gilio-Whitaker, *As Long as the Grass Grows*, 153.

23. "Recycling Makes a Difference in Akiak, Alaska," Environmental Protection Agency, Region 10, December 2015, https://www.epa.gov/sites/production/files/2016-01/documents/r10-igap-success-story-akiak.pdf.

24. Gilio-Whitaker, *As Long as the Grass Grows*, 132–35.

25. Gilio-Whitaker, *As Long as the Grass Grows*, 149.

26. Gilio-Whitaker, *As Long as the Grass Grows*, 152–54.

27. Gilio-Whitaker, *As Long as the Grass Grows*, 142.

28. EJ Practitioners Network, "As Long as Grass Grows: A Conversation with Author Dina Gilio-Whitaker," YouTube video, 44:57, May 24, 2019, https://www.youtube.com/watch?v=l6q1QLu1RWQ&t=205s.

29. Dina Gilio-Whitaker, *As Long as the Grass Grows*, 162.

30. Casey Parks, "Cully Park Will Get $125 Million Boost from Portland Parks and Recreation, City Commissioner Amanda Fritz Says," *Oregonian*, February 12, 2014, http://www.oregonlive.com/portland/index.ssf/2014/02/cully_park_will_get_125_millio.html.

31. "FSC Group Certification," Sustainable Northwest, accessed January 18, 2016, http://www.sustainablenorthwest.org/what-we-do/programs/FSC-group-certification.

32. Jon Biemer et al., "Environmental Justice at Wy'east," *Wy'east Beacon*, July/August 2015, http://wyeastuu.org/files/newsletter/July_August_wyeast_2015.pdf.

33. Home page, *Censored News*, accessed April 21, 2020, https://bsnorrell.blogspot.com/.

34. "Our Story," Deep South Center for Environmental Justice, accessed August 23, 2019, http://www.dscej.org/our-story.

35. "2021 Policy Agenda," We Act for Environmental Justice, accessed December 14, 2020, https://www.weact.org/wp-content/uploads/2020/12/2021-Policy-Agenda-ENG-FINAL.pdf.

36. Home page, Big Elk Native American Center, accessed November 1, 2020, http://www.bigelknativecenter.org/; Pappan, John (acting director of Big Elk Native American Center), emails to Jon Biemer, October 13 and 14, 2020.

37. "Mapping Ecological Stewardship Opportunities," First Nations Development Institute, accessed October 23, 2020, https://www.firstnations.org/projects/mapping-ecological-stewardship-opportunities-meso/.

38. Home page, Honor the Earth, accessed March 20, 2020, http://www.honorearth.org/.

39. Home page, Indian Law Center, accessed March 20, 2020, https://indianlaw.org/issue/environmental-protection.

40. "Our Work," Movement Rights, accessed March 20, 2020, https://www.movementrights.org/our-work/.

41. "About Us," Native American College Fund, accessed May 16, 2020, https://collegefund.org/about-us/.

42. Home page, Red Cloud Renewable Energy Center, accessed April 6, 2020, https://www.redcloudrenewable.org/; Richard Fox, RCREC communications director, personal email to author, March 28, 2020.

CHAPTER 7

1. *Transforming Our World: The 2030 Agenda for Sustainable Development*, United Nations General Assembly, adopted September 25, 2015, http://www.un.org/ga/search/view_doc.asp?symbol=A/RES/70/1&Lang=E.

2. "The Story Behind the Goals," 17Goals, accessed February 16, 2020, http://17goals.org/the-story-behind-the-goals.

3. "SDG Acceleration Actions," Sustainable Development Goals Knowledge Platform, accessed February 19, 2020, https://sustainabledevelopment.un.org/sdgactions.

4. Shamini Dhana, "Fair Trade and Fairtrade. What's the Difference and Why Does It Matter?" Shamini Dhana, October 9, 2015, http://www.shaminidhana.com/journal/fair-trade-vs-fairtrade-and-why-it-matters.

5. "Coffee," Fairtrade America, accessed November 6, 2019, http://www.fairtradeamerica.org/Fairtrade-Products/Coffee.

6. "Economic Development," Heifer International, accessed August 31, 2019, https://www.heifer.org/our-work/economic-development/index.html.

7. Pat Morrison, "Lender 2.0," *Los Angeles Times*, May 14, 2011, http://articles.latimes.com/2011/may/14/opinion/la-oe-morrison-premal-shah-043011.

8. "UpEnergy," Kiva, accessed January 26, 2016, http://www.kiva.org/partners/290.

9. "Hasan," Kiva, accessed February 27, 2015, http://www.kiva.org/lend/844846.

10. "Population Ages 0–14 (Percent of Total)," World Bank, accessed February 16, 2017, http://data.worldbank.org/indicator/SP.POP.0014.TO.ZS.

11. Carolee Buckler and Heather Creech, *Shaping the Future We Want: UN Decade for Education for Sustainable Development (2005–2014), Final Report*, United Nations Scientific, Educational, and Cultural Organization, 2014, http://unesdoc.unesco.org/images/0023/002301/230171e.pdf.

12. Buckler and Creech, *Shaping the Future*, 66.

13. Rosalyn McKeown, *Education for Sustainable Development Toolkit, Version 2.0* (self-pub., July 2002), http://www.esdtoolkit.org.

14. Buckler and Creech, *Shaping the Future*, 50.

15. Buckler and Creech, *Shaping the Future*, 52.

16. Buckler and Creech, *Shaping the Future*, 55.

17. Buckler and Creech, *Shaping the Future*, 59.

18. "Global RCE Network," United Nations University, Institute for the Advanced Study of Sustainability, accessed February 19, 2020, http://www.rcenetwork.org/portal/rces-worldwide.

19. "Afghanistan," Girls Not Brides, accessed October 22, 2020, https://www.girlsnotbrides.org/child-marriage/afghanistan.

20. Home page, Central Asia Institute, accessed March 9, 2020, https://centralasiainstitute.org.

21. "Impact," Growth Through Learning, accessed October 22, 2020, https://growththroughlearning.org/impact-2.

22. "Empowering Women and Girls through Education," Central Asia Institute, accessed March 20, 2018, https://centralasiainstitute.org/why-it-matters.

23. Elina Pradhan, "Female Education and Childbearing: A Closer Look at the Data," World Bank Blogs, November 24, 2015, https://blogs.worldbank.org/health/female-education-and-childbearing-closer-look-data.

24. "Human Population: Women," Population Reference Bureau, accessed February 12, 2012, http://www.prb.org/Publications/Lesson-Plans/HumanPopulation/Women.aspx.

25. "Human Population."

26. "Mujeres indigenas in America Latina," Centro Latinamericano y Caranino de Demographica (CELACD)—Division de Pobplacion y Division de Asuntos de Genaro de la CESPAL, 2013, Cuadro 12, 71.

27. "Education and Population Dynamics: Mobilizing Minds for a Sustainable Future," United Nations Educational, Scientific and Cultural Organization, EPD-99/WS/1, March 1999, http://unesdoc.unesco.org/images/0011/001163/116355eo.pdf.

28. Liz Neporent, "7 Facts about Condoms," *ABC News*, February 8, 2011, https://abcnews.go.com/Health/condom-facts-things/story?id=12869404.

29. "2018 Annual Report," Friends of UNFPA, accessed November 19, 2019, https://www.friendsofunfpa.org/wp-content/uploads/2019/11/Friends-of-UN FPA-Annual-Report-2018-1.pdf.

30. "Paul Hawken, "Drawdown: The Most Comprehensive Plan Ever Proposed to Reverse Global Warming," YouTube video, 1:17, KODX Seattle, April 21, 2017, https://www.youtube.com/watch?v=KYvKv0lM-_A.

31. "Shop with Intention; Share the Joy," Ten Thousand Villages, accessed December 13, 2019, https://www.tenthousandvillages.com/about-us.

32. "Products and Partners," Fair Trade USA, accessed December 25, 2016, http://www.fairtradeusa.org/products-partners.

33. Home page, Bedre Fine Chocolate, accessed October 31, 2020, https://bedrechocolates.com.

34. Home page, Beyond Buckskin Boutique, accessed October 31, 2020, https://shop.beyondbuckskin.com.

35. Home page, Eighth Generation, accessed October 31, 2020, https://eight hgeneration.com.

36. Home page, Native Harvest, accessed October 31, 2020, https://nativehar vest.com.

37. "Environmental Sustainability," Heifer International, accessed September 1, 2019, https://www.heifer.org/our-work/flagship-projects/index.html.

CHAPTER 8

1. "'Founder of Environmental Education,' Has Died," University of Michigan, accessed October 23, 2020, http://ns.umich.edu/Releases/2001/May01/r053101b.html.

2. *K–12 Environmental Education: Guidelines for Excellence*, North American Association for Environmental Education, 2019, https://naaee.org/eepro/publication/excellence-environmental-education-guidelines-learning-k-12.

3. "CPSC Elementary School," Central Park School for Children, accessed October 23, 2020, http://cpscnc.org/CPSCElementary.

4. "Curriculum," Pennsylvania Department of Education, accessed October 23, 2020, https://www.education.pa.gov/Teachers%20-%20Administrators/Cur riculum/Pages/default.aspx.

5. "Keystone Exams: Biology," Pennsylvania Department of Education, accessed September 30, 2017, http://static.pdesas.org/Content/Documents/Biol ogy%20Keystone%20Assessment%20Anchors%20and%20Eligible%20Content%20 April%202014.pdf.

6. "Wisconsin's Environmental and Conservation History," Environmental Education in Wisconsin, Wisconsin Green Schools Network, accessed September 30, 2017, http://eeinwisconsin.org/resource/about.aspx?s=123926.0.0.2209; Home

page, Kentucky Green and Healthy Schools, accessed April 21, 2020, https://greenschools.ky.gov/Pages/default.aspx.

7. Home page, Ecology in Classrooms and Outdoors, accessed October 23, 2020, https://www.ecologyoutdoors.org/.

8. "Institution Status," Association of Zoos and Aquariums, accessed November 13, 2019, https://www.aza.org/inst-status.

9. "Mariana Avifauna Conservation Project," Toledo Zoo, accessed October 23, 2020, http://www.pacificbirdconservation.org/mariana-avifauna-conservation-program.html.

10. "Zoos and Aquariums," World Conservation Society, accessed December 30, 2015, http://www.wcs.org/parks.

11. "2020 Strategy," World Conservation Society, accessed December 30, 2015, http://www.wcs.org/our-work/2020-strategy.

12. "Teacher Programs," Monterey Aquarium, accessed May 30, 2020, https://www.montereybayaquarium.org/for-educators/teacher-professional-development/teacher-programs.

13. "SAFE Saving Animals from Extinction," Association of Zoos and Aquariums, accessed November 13, 2019, https://www.aza.org/safe-species.

14. Pattie Ensel Bailie, "Forest School in Public School: Is It Possible?" Natural Start Alliance, April 2014, http://naturalstart.org/feature-stories/forest-school-public-school-it-possible.

15. Home page, American Forest Kindergarten Association, accessed October 23, 2020, https://www.forestkindergartenassociation.org/.

16. Bethany Barnes, "Oregon Ballot Measures: Measures 98 and 99 Pass," *Oregonian*, November 8, 2016, http://www.oregonlive.com/politics/index.ssf/2016/11/oregon_ballot_measures_measure.html.

17. Tom Potter and Karen Hansen, "My View: Oregon Outdoor School: A Legacy Worth Saving," *Portland Tribune*, June 16, 2015, http://www.pamplinmedia.com/pt/10-opinion/263732-135860-my-view-outdoor-school-a-legacy-worth-saving.

18. Associated Press, "Oregon Decides Whether to Fund an Outdoor School Program," *Fox News U.S.*, October 26, 2016, https://www.foxnews.com/us/oregon-decides-on-whether-to-fund-an-outdoor-school-program.

19. Home page, Trackers School, accessed July 23, 2014, https://www.trackerschool.com/.

20. Home page, Deerdance, accessed December 29, 2015, http://www.deerdance.org/.

21. "About Us," Wilderness Awareness School, accessed December 28, 2015, http://wildernessawareness.org/about/about.

22. "Our Mission," Outward Bound, accessed December 28, 2015, http://www.outwardbound.org/about-outward-bound/outward-bound-today/.

23. "Find Your Course," National Outdoor Leadership School, accessed December 29, 2015, http://www.nols.edu/courses/; Wikipedia, s.v. "National Out-

door Leadership School," accessed October 23, 2020, https://en.wikipedia.org/wiki/National_Outdoor_Leadership_School.

24. "Courses Offered: Elective Courses Available for the ELS Degree," Arizona State University, accessed January 2, 2016, http://els.asu.edu/curriculum/courses offered.

25. "Degree Requirements," College of the Atlantic, accessed October 23, 2020, https://www.coa.edu/academics/human-ecology-degree/degree-require ments/.

26. William Dietrich, *The Final Forest: The Battle for the Last Great Trees of the Pacific Northwest* (New York: Penguin Books, 1992), 47–55.

27. "About Us," University of Oregon Environmental Studies Program, accessed January 2, 2016, http://envs.uoregon.edu/reference/.

28. "Did You Know . . . ? Marine Life / Ocean Facts . . . " MarineBio, accessed December 13, 2019, https://marinebio.org/creatures/facts/.

29. "Scripps Education," Scripps Institution of Oceanography, accessed December 13, 2019, https://scripps.ucsd.edu/education.

30. "Why Study Ocean Sciences at Dal?" Dalhousie University, accessed December 13, 2019, https://www.dal.ca/academics/programs/undergraduate/ocean-sciences.html; "Ocean Tracking Network in Brief," Ocean Tracking Network, accessed December 13, 2019, http://oceantrackingnetwork.org/about/#brief.

31. "Green and Healthy Schools Academy," Green Building Alliance, accessed October 23, 2020, https://www.go-gba.org/initiatives/green-healthy-schools-academy/.

32. "About Us," Association of Zoos and Aquariums, accessed December 15, 2016, https://www.aza.org/about-us.

33. "SARE Nationwide," Sustainable Agriculture Research and Education, accessed December 15, 2016, http://www.sare.org/.

34. "A Guide to Green Careers," Learn How to Become, accessed November 14, 2019, https://www.learnhowtobecome.org/career-resource-center/green-careers/.

35. Andrejs Kulnieks, Dan Roronhiakewen Longboat, and K. Young, eds., *Contemporary Studies in Environmental and Indigenous Pedagogies* (Switzerland: Springer Nature, 2013), overview accessed February 2, 2017, https://www.springer.com/us/book/9789462092938.

CHAPTER 9

1. Richard Louv, *Last Child in the Woods* (Chapel Hill, NC: Algonquin Books, 2008).

2. Maria Alice Campos Friere, "Our Sacred Planet, Our Sacred Mother," (presentation), Natural Way series, Earth and Spirit Council, Portland, Oregon, November 23, 2010.

3. "Press: Benefits of Camp," American Camp Association, accessed December 28, 2015, http://www.acacamps.org/press-room/benefits-of-camp.

4. "YMCA Camp Lakewood," Camp Channel, accessed December 28, 2015, http://www.campchannel.com/summer-camps/YMCA-Camp-Lakewood-2413.html.

5. "Summer Camping Program," The Fresh Air Fund, accessed November 14, 2019, http://www.freshair.org/summer-camping-program.

6. In February 2020, the Boy Scouts of America filed bankruptcy to settle sexual abuse claims and reorganize. I had a positive scouting experience.

7. "Environmental Education," City of Portland Parks and Recreation, accessed July 26, 2014, www.portlandoregon.gov/parks/38295.

8. "Number of Participants in Hiking in the United States from 2006 to 2017," Statista, accessed September 3, 2019, https://www.google.com/search?q=how+many+people+go+hiking&oq=how+many+people+go+hiking+&aqs=chrome..69i57j69i64.9695j0j7&sourceid=chrome&ie=UTF-8; "Public Data: Population: United States," Google, accessed September 5, 2019, https://www.google.com/publicdata/explore?ds=kf7tgg1uo9ude_&met_y=population&idim=country:US&hl=en&dl=en. (Based on US Census Bureau data, which did not explicitly present 2017 population.)

9. "Our Story," International Ecotourism Society, accessed January 14, 2016, https://www.ecotourism.org/our-story.

10. "Project Summaries," International Ecotourism Society, accessed January 6, 2016, https://www.ecotourism.org/project-summaries.

11. "National Parks and Reserves," Costa Rica Guide, accessed October 23, 2020, https://costa-rica-guide.com/nature/parks-reserves/.

12. Pamela K. Brodowsky, *Ecotourists Save the World* (New York: Penguin, 2010).

13. Wikipedia, s.v. "John James Audubon," accessed December 5, 2014, https://en.wikipedia.org/wiki/John_James_Audubon.

14. "John Muir Biography," *Biography*, accessed January 28, 2016, http://www.biography.com/people/john-muir-9417625.

15. Douglas Brinkley, *The Quiet World, Saving Alaska's Wilderness Kingdom, 1879–1960* (New York: HarperCollins, 2011), 16, 169.

16. Aldo Leopold, *A Sand County Almanac, with Essays on Conservation from Round River* (New York: Ballantine Press, 1974), 205.

17. "Citizen Science," Cornell University, accessed January 2, 2016, http://www.birds.cornell.edu/page.aspx?pid=1664.

18. Wikipedia, s.v. "Amory Lovins," accessed December 6, 2014, http://en.wikipedia.org/wiki/Amory_Lovins.

19. "Old World New," Old World New, accessed November 15, 2019, https://oldworldnew.us/.

20. Home page, Thinking Country, accessed November 15, 2019, https://thinkingcountry.com/.

21. Home page, Mary Ellen Hannibal, accessed September 3, 2017, https://www.maryellenhannibal.com/.

22. "Ensia Mentor Program," Ensia, accessed February 2, 2017, https://ensia.com/about/mentor-program/.

23. "Mentor Program," Society of Environmental Journalists, accessed February 2, 2017, http://www.sej.org/initiatives/mentor-program/overview.

24. "Waste and Response, Mentors," Institute for Tribal Environmental Professionals, Northern Arizona University, accessed February 2, 2017, https://www7.nau.edu/itep/main/Waste/waste_mentors.

25. Home page, Think Beyond Plastics, accessed October 23, 2020, https://www.thinkbeyondplastic.com/.

CHAPTER 10

1. "The Issues," The Land Institute, accessed September 27, 2014, http://www.landinstitute.org/our-work/issues/.

2. James MacDonald and Robert Hoppe, "Examining Consolidation in U.S. Agriculture," Department of Agriculture, Economic Research Service, March 14, 2018, https://www.ers.usda.gov/amber-waves/2018/march/examining-consoli dation-in-us-agriculture/.

3. James MacDonald, Penni Korb, and Robert Hoppe, "Farm Size and the Organization of US Crop Farming," Department of Agriculture, Economic Research Report no. 152, August 2013, https://www.ers.usda.gov/webdocs/publica tions/45108/39359_err152.pdf.

4. Emily Monaco, "Certified Organic Acreage on the Rise," *Organic Authority*, updated May 7, 2019, https://www.organicauthority.com/buzz-news/certified-organic-acreage-on-the-rise.

5. "2014 and 2015 Organic Certifier Data," National Agricultural Statistics Service, U.S. Department of Agriculture, accessed April 12, 2017, https://www.nass.usda.gov/Surveys/Guide_to_NASS_Surveys/Organic_Production/Or ganic_Certifiers/2016/USDA_Accredited_Certifying_Agent_Certified_Organic_Data_2014_2015.pdf; Monaco, "Certified Organic Acreage on the Rise."

6. "What Is Anthroposophy?" Waldorf Answers, accessed August 13, 2015, http://www.waldorfanswers.com/Anthroposophy.htm.

7. Wikipedia, s.v. "Walter James, Fourth Baron Northbourne," accessed August 14, 2015, https://en.wikipedia.org/wiki/Walter_James,_4th_Baron_Northbourne. Wikipedia, s.v. "Biodynamic Agriculture," accessed August 14, 2015, https://en.wikipedia.org/wiki/Biodynamic_agriculture.

8. "Part 205—National Organic Program," Electronic Code of Federal Regulations, Title 7, Subtitle B, Chapter 1, Subchapter M, Part 205, Paragraphs 205.202, 205.203, 205.239, 205.271, 205.403, and 205.130, current as of April 22,

2020, https://www.ecfr.gov/cgi-bin/text-idx?tpl=/ecfrbrowse/Title07/7cfr205_main_02.tpl.

9. Kate Medley, "Tuscarora Growers," Whole Foods, August 10, 2011, http://www.wholefoodsmarket.com/blog/whole-story/tuscarora-organic-growers.

10. "Organic Matters," Earthbound Farm, accessed October 23, 2020, https://www.earthboundfarm.com/our-commitment/organic/.

11. "Supporting Our Farmers," Horizon Organic, accessed October 24, 2020, https://horizon.com/our-farmers/supporting-our-farmers/.

12. M. Shahbandeh, ed., "Organic Food and Non-food Sales in the United States in 2013," Statista; M. Shahbandeh, ed., "Organic Food and Non-food Sales in the United States in 2018," Statista, http://www.statista.com/statistics/244394/organic-sales-in-the-united-states/.

13. Jonathon Knudson, "U.S. Organic Market Tops $50 Billion," *Agweek*, June 10, 2019, https://www.agweek.com/business/agriculture/4622665-us-organic-market-tops-50-billion.

14. Maggie McNeil (contact), "COVID-19 Will Shape Organic Industry in 2020 after Banner Year in 2019," Organic Trade Association, June 9, 2020, https://ota.com/news/press-releases/21328.

15. Kristin Ohlson, *The Soil Will Save Us: How Scientists, Farmers, and Foodies Are Healing the Soil to Save the Planet* (New York: Rodale, 2014), 232.

16. Ohlson, *The Soil Will Save Us*, 231.

17. Ohlson, *The Soil Will Save Us*, 42.

18. Eric Toensmeier, *The Carbon Farming Solution* (White River Junction, VT: Chelsea Green, 2016), 12.

19. Judith Schwartz, *Cows Save the Planet* (White River Junction, VT: Chelsea Green, 2013), 24.

20. Toensmeier, *Carbon Farming Solution*, 54–55.

21. Ohlson, *The Soil Will Save Us*, 137.

22. Ohlson, *The Soil Will Save Us*, 140.

23. Ohlson, *The Soil Will Save Us*, 93.

24. Sheryl Karas, "Australian Soil Ecologist Christine Jones Gives a Paradigm-Changing Workshop at Chico State," California State University Chico, based on workshop given June 24, 2019, https://www.csuchico.edu/regenerativeagriculture/blog/christine-jones-event.shtml.

25. "Articles," Amazing Carbon, accessed April 24, 2020, https://www.amazingcarbon.com/.

26. Home page, Regenerative Organic Alliance, accessed August 12, 2018, https://regenorganic.org/. "Framework for Regenerative Organic Certification," Regenerative Organic Alliance, May 2018, https://regenorganic.org/wp-content/uploads/2018/06/ROC-Framework-1-May-2018.pdf.

27. Toensmeier, *Carbon Farming Solution*, 45–46.

28. Toensmeier, *Carbon Farming Solution*, 121.

29. Toensmeier, *Carbon Farming Solution*, 70.

30. "Science," The Land Institute, accessed September 27, 2014, http://www.landinstitute.org/our-work/science/. Toensmeier, *Carbon Farming Solution*, 121–22.

31. Allan Savory, "How to Fight Desertification and Reverse Climate Change," TED video, 22:13, February 2013, https://www.ted.com/talks/allan_savory_how_to_green_the_world_s_deserts_and_reverse_climate_change.

32. Judith Schwartz, "Cows Save the Planet, and Other Improbable Ways of Restoring Soil to Heal the Earth," Judith Schwartz (blog), accessed August 9, 2015, http://www.judithdschwartz.com/.

33. "Joel's Bio," Polyface Farm, accessed September 27, 2014, http://www.polyfacefarms.com/speaking-protocol/joels-bio/.

34. Toensmeier, *Carbon Farming Solution*, 40.

35. Wikipedia, s.v. "*Leucaena leucocephala*," accessed July 12, 2017, https://en.wikipedia.org/wiki/Leucaena_leucocephala.

36. Toensmeier, *Carbon Farming Solution*, 96.

37. Toensmeier, *Carbon Farming Solution*, 42–43.

38. Teresa Koper, "FAQ: Agricultural Carbon Credit Projects," Climate Trust, accessed October 14, 2014, https://www.climatetrust.org/faq-agricultural-carbon-credit-projects/.

39. Ohlson, *The Soil Will Save Us*, 207

40. Home page, Sustainable Agriculture Research and Education, accessed December 12, 2016, http://www.sare.org/.

CHAPTER 11

1. "Garden to Table: A Five-Year Look at Gardening in America," National Gardening Association, 2014, https://garden.org/special/pdf/2014-NGA-Garden-to-Table.pdf.

2. "Here's the Dirt on City Parks: Community Gardens Are Growing," Trust for Public Lands, August 22, 2018, https://www.tpl.org/blog/here%E2%80%99s-dirt-park-trends-community-gardens-are-growing.

3. "Resources," American Community Gardening Association, accessed October 24, 2020, https://www.communitygarden.org/resources/.

4. Arthur Neal, "Add Your Market to the 2011 USDA National Farmer's Market Directory Today!" U.S. Department of Agriculture, February 21, 2017, https://www.usda.gov/media/blog/2011/04/28/add-your-market-2011-usda-national-farmers-market-directory-today (1994 count); "Local Food Directory: Farmers Market Directory," U.S. Department of Agriculture, updated December 11, 2019, https://www.ams.usda.gov/local-food-directories/farmersmarkets (2019 count).

5. "What Is Biodynamics?" Biodynamic Association, accessed August 14, 2015, https://www.biodynamics.com/what-is-biodynamics.

6. Steven McFadden, "The History of Community Supported Agriculture, Part I: Community Farms in the 21st Century: Poised for Another Wave of Growth?" Rodale Institute, accessed February 6, 2020, https://web.archive.org/web/20121127113607/http://newfarm.rodaleinstitute.org/features/0104/csa-history/part1.shtml.

7. Katherine Adam, "Community Supported Agriculture," National Center for Appropriate Technology, accessed September 25, 2014, https://attra.ncat.org/attra-pub/viewhtml.php?id=262.

8. "CSA Database," Wilson College, accessed September 25, 2014; "Robyn Van En Center," Wilson College, accessed July 27, 2019, https://www.wilson.edu/robyn-van-en-center.

9. "About Us," LocalHarvest, accessed September 25, 2014, http://www.localharvest.org/about.jsp.

10. Alex Steffen, ed., *WorldChanging: A User's Guide for the 21st Century* (New York: Abrams, 2011), 517.

11. Wikipedia, s.v. "Fair Packaging and Labeling Act," accessed September 17, 2014, http://en.wikipedia.org/wiki/Fair_Packaging_and_Labeling_Act.

12. Wikipedia, s.v. "Organic Certification," accessed September 17, 2014, http://en.wikipedia.org/wiki/Organic_certification.

13. Allie Morris, "For Some Small Farmers in New Hampshire Certified Naturally Grown Makes More Sense," *Concord Monitor*, August 4, 2014, http://www.concordmonitor.com/community/town-by-town/weare/12980178-95/for-some-small-farmers-in-new-hampshire-certified-naturally-grown-makes-more-sense.

14. Home page, Montana Sustainable Growers Union, accessed June 12, 2015, http://www.homegrownmontana.org/.

15. "Transitioning Farmland to Organic," Oregon Tilth, accessed October 24, 2020, https://tilth.org/knowledgebase_category/transition/#transitional-certification.

16. "History," Non-GMO Project, accessed October 24, 2020, https://www.nongmoproject.org/about/history/. Home page, Non-GMO Project, accessed September 11, 2019, https://livingnongmo.org/.

17. "Vermont Public Interest Group and Center for Food Safety Move to Defend Vermont GE Labeling Law," Center for Food Safety, July 21, 2014, http://www.centerforfoodsafety.org/issues/976/ge-food-labeling/international-labeling-laws.

18. "International Labeling Laws," Center for Food Safety, September 17, 2014, http://www.centerforfoodsafety.org/issues/976/ge-food-labeling/international-labeling-laws.

19. Lucas Reijnders and Sam Soret, "Quantification of Environmental Impacts of Different Dietary Food Choices," *American Journal of Clinical Nutrition*, 2003, http://ajcn.nutrition.org/content/78/3/664S.full.

20. Home page, Green Smoothie Girl, accessed June 12, 2015, http://greens moothiegirl.com/.

21. Sandor Katz, ed., *The Art of Fermentation* (White River Junction, VT: Chelsea Green, 2012).

22. Sandor Katz, ed., "Welcome to Wild Fermentation Portal," Wild Fermentation, accessed September 29, 2014, http://www.wildfermentation.com/; "Discover the Digestive Benefits of Fermented Foods," *Health & Nutrition Newsletter*, Tufts University, February 2014, http://www.nutritionletter.tufts.edu/issues/10_2/cur rent-articles/Discover-the-Digestive-Benefits-of-Fermented-Foods_1383-1.html.

23. "Retailers Wanted for National Bulk Foods Week," *Progressive Grocer*, July 11, 2014, http://www.progressivegrocer.com/industry-news-trends/national-supermarket-chains/retailers-wanted-national-bulk-foods-week.

24. Maura Judkis, "You Might Think There Are More Vegetarians Than Ever. You'd Be Wrong," *Washington Post*, August 3, 2018, https://www.washington post.com/news/food/wp/2018/08/03/you-might-think-there-are-more-vegetar ians-than-ever-youd-be-wrong/.

25. Kari Hamerschlag, "Meat Eaters Guide: Report," Environmental Working Group, accessed October 27, 2017, https://www.ewg.org/meateatersguide/a-meat-eaters-guide-to-climate-change-health-what-you-eat-matters/climate-and-environmental-impacts/.

26. Wikipedia, s.v., "Atmospheric Methane," edited October 1, 2020, https://en.wikipedia.org/wiki/Atmospheric_methane.

27. Millie Milliken, "Goat: The Other (Sustainable) Red Meat," *Triple Pundit*, Presidio Graduate School, Fall 2011, http://www.triplepundit.com/2011/12/goat-other-sustainable-red-meat/.

28. Sami Grover, "Goat Meat as an Ethical Alternative to Beef," *Tree Hugger*, April 11, 2011, http://www.treehugger.com/green-food/goat-meat-as-an-ethical-alternative-to-beef.html.

29. United Nations, *Greenhouse Gas Emissions from Ruminant Supply Chains* (Rome: United Nations, Food and Agriculture Organization of the United Nations, 2013), 37, http://www.fao.org/docrep/018/i3461e/i3461e00.htm.

30. Nina Rastogi, "The Kindest Cut, Which Cut of Meat Harms Our Planet the Least?" *Slate*, April 28, 2009, http://www.slate.com/articles/health_and_sci ence/the_green_lantern/2009/04/the_kindest_cut.html.

31. Rastogi, "Kindest Cut."

32. "About Chickens Farmed for Meat," Compassion in World Farming, accessed July 25, 2016, http://www.ciwf.org.uk/farm-animals/chickens/meat-chickens/.

33. Anita Wolff, "The Pros and Cons of Fish Farming," *Encyclopedia Britannica Advocacy for Animals*, August 4, 2008, http://advocacy.britannica.com/blog/advo cacy/2008/08/the-pros-and-cons-of-fish-farming/.

34. "New Seasons Expands Green Wheels to All Locations," *Progressive Grocer*, April 18, 2016, https://progressivegrocer.com/new-seasons-expands-green-wheels-all-locations#close-olyticsmodal.

35. "FAQs," Trader Joe's, accessed September 27, 2019, https://www.trader joes.com/faqs/product-information.

36. Bruce Lee, "What Trader Joe's Will Do about the Chemicals on Your Receipts," *Forbes*, January 21, 2018, https://www.forbes.com/sites/brucelee/2018/01/21/what-trader-joes-will-do-about-the-chemicals-on-your-receipts/#4978503725f5.

37. "Community Supported Agriculture," Local Harvest, accessed November 22, 2019, https://www.localharvest.org/csa/.

38. "Falling Fruit—Map the Harvest," Falling Fruit, accessed October 31, 2019, http://www.fallingfruit.org/.

39. Elizabeth Leader Smith, "Biodynamic 101: What to Know and Buy Now," Whole Foods Market, April 16, 2018, https://www.wholefoodsmarket.com/blog/biodynamic-101.

40. Dawn Gifford, "20 Perennial Plants to Plant Once and Enjoy Forever," Small Footprint Family, accessed September 3, 2017, https://www.smallfootprint family.com/20-perennial-vegetables.

41. "Buying Plant-Based Alternatives," Compassion in World Farming, accessed November 22, 2019, https://www.ciwf.com/your-food/plant-based-alternatives/.

42. "Diet Tips," Eat Low Carbon, accessed August 24, 2018, http://eatlowcar bon.org/diet-tips/.

CHAPTER 12

1. "Less Is More: The Tiny House Movement," CustomMade, August 8, 2014, http://www.custommade.com/blog/tiny-house-movement/.

2. "Designers and Builders," Small Home Society, accessed September 9, 2015, http://smallhousesociety.net/designers-builders/.

3. D&R International, *2011 Building Energy Data Book* (Richland, WA: Pacific Northwest National Laboratory, updated March 2012), 1–30, Table 1.5.4, https://ieer.org/wp/wp-content/uploads/2012/03/DOE-2011-Buildings-Energy-Data-Book-BEDB.pdf.

4. *2015 Residential Energy Consumption Survey (RECS)*, U.S. Energy Information Agency, Table CE1.1, released May 2018, https://www.eia.gov/consumption/residential/data/2015/c&e/pdf/ce1.1.pdf.

5. David Owen, "The Greenest Place in the U.S. May Not Be Where You Think," *Environment 360*, October 26, 2009, http://e360.yale.edu/feature/green est_place_in_the_us_its_not_where_you_think/2203/.

6. Megan Rosa, "Looking Back: LEED History," Sustainable Investment Group, August 8, 2016, https://sigearth.com/leed-history/.

7. "King County Signature Report, June 24, 2008, Ordinance 16147," King County, Washington, https://kingcounty.gov/~/media/depts/dnrp/solid-waste/green-building/documents/KC-green-building-ordinance.ashx?la=en.

8. "LEEDv4," US Green Building Council, accessed October 25, 2020, https://www.usgbc.org/leed/v4.

9. "LEED and the International Construction Code," US Green Building Council, November 2018, https://www.usgbc.org/sites/default/files/2018%20IgCC%20Policy%20Brief.pdf.

10. Heather Benjamin, "LEED Zero Petinelli Headquarters Creates Momentum for Client Goals," U.S. Green Building Council, April 25, 2019, https://www.usgbc.org/articles/leed-zero-petinelli-headquarters-creates-momentum-client-goals.

11. "LEED Zero Program Guide," U.S. Green Building Council, updated May 2019, https://www.usgbc.org/sites/default/files/LEED_Zero_Program_Guide.pdf.

12. T. Wang, "Green Buildings in the U.S.—Statistics and Facts," Statista, accessed August 28, 2019, https://www.statista.com/topics/1169/green-buildings-in-the-us/.

13. "LEED Professional Credentials," U.S. Green Building Council, accessed November 9, 2019, https://new.usgbc.org/credentials.

14. "Shop at Habitat for Humanity ReStore," Habitat for Humanity, accessed October 25, 2020, https://www.habitat.org/restores/shop.

15. Richard Florida, "Why Bigger Cities Are Greener," City Lab, April 19, 2012, http://www.citylab.com/work/2012/04/why-bigger-cities-are-greener/863/.

16. Windy A. Jordon, "Deconstruction Can Be a Tax-Savvy Alternative to Demolition," *Washington Post*, August 25, 2016, https://www.washingtonpost.com/realestate/deconstruction-can-provide-huge-tax-benefits-for-property-owners/2016/08/24/8f6c5270-62fb-11e6-96c0-37533479f3f5_story.html?noredirect=on.

17. "The ReUse People of America," The ReUse People of America, accessed August 16, 2019, http://thereusepeople.org/findus.

18. "Programs," DESIRE, North Carolina Energy Technology Center, accessed January 4, 2021, https://programs.dsireusa.org/system/program.

19. "Best Management Practices for Solar Installation Policy, Planning Improvements," National Association of Regional Councils, accessed December 15, 2016, http://narc.org/wp-content/uploads/BMP-Planning-Step-2-PL-2-A-Adopt-a-Solar-Access-Ordinance_FINAL.pdf.

20. "City of Kansas City—Solar and Geothermal Access," Database for State Incentives for Renewables and Efficiency (DSIRE), North Carolina Clean Energy Technology Center, updated March 17, 2016, http://programs.dsireusa.org/system/program/detail/5581.

CHAPTER 13

1. "How Much Do People Walk in Their Lifetime?" *WiseGeek*, accessed August 20, 2015, http://www.wisegeek.com/how-much-do-people-walk.htm. The author refers to The Cleveland Clinic.

2. Vinod Panchbhavi, "Foot Bone Anatomy," *Medscape*, March 8, 2013, http://emedicine.medscape.com/article/1922965-overview.

3. Erica Chu and Donald Resnick, "Sesamoid Bones, Normal and Abnormal," *Radsource*, MRI Web Clinic, June 2014, http://radsource.us/sesamoid-bones/.

4. Richard Florida, "The Great Divide in How Americans Commute to Work," CityLab, January 22, 2019, https://www.citylab.com/transportation/2019/01/commuting-to-work-data-car-public-transit-bike/580507/.

5. Caroline Criado Perez, *Invisible Women: Data Bias in a World Designed for Men* (New York: Abrams, 2019), 34.

6. E. Mazareaunu, "Gender Distribution of Public Transit Travelers in the United States from 2008 to 2015, by Transit Mode," Statista, January 5, 2018, https://www.statista.com/statistics/715212/public-transit-use-gender-transit-mode-united-states/.

7. "Human Power," The Exploratorium, accessed August 20, 2015, https://www.exploratorium.edu/cycling/humanpower1.html.

8. "Development of the Bicycle," America on the Move, *Smithsonian*, accessed August 20, 2015, http://amhistory.si.edu/onthemove/themes/story_69_2.html.

9. Bill Nye, *Unstoppable, Harnessing Science to Change the World* (New York: St. Martin's Press, 2015), 185.

10. "Bicycling and Walking in the United States: 2018 Benchmarking Report," League of American Bicyclists, February 5, 2019, https://bikeleague.org/content/bicycling-and-walking-united-states-2018-benchmarking-report.

11. *The Role of Walking and Cycling in Reducing Congestion, A Portfolio of Measures* (Brussels: FLOW Project, July 2016), 61, http://h2020-flow.eu/uploads/tx_news/FLOW_REPORT_-_Portfolio_of_Measures_v_06_web.pdf; Patrick Sisson, "Sick of Traffic? Get Cities to Invest in Bikes and Mass Transit, Says Report," *Curbed*, June 11, 2019, https://www.curbed.com/2019/6/11/18661586/bike-train-traffic-transportation-congestion.

12. Brad Auerbach, "Electric Bikes Are Shaping the Future—and This Company Already Has," *Forbes*, January 16, 2017, https://www.forbes.com/sites/bradauerbach/2017/01/16/electric-bikes-will-shape-the-future-here-is-a-company-already-underway.

13. "800 mpg E-bike," Alan's E-Bikes, accessed October 25, 2020, http://www.alansebikes.com/electric-bike-mileage.

14. Al Gore, *Our Choice: A Plan to Solve the Climate Crisis* (Emmaus, PA: Rodale, 2009), 269.

15. Michael J. Coren, "Electric Cars Are Changing the Cost of Driving," *Quartz*, November 8, 2019, https://qz.com/1737145/the-economics-of-driving-seven-teslas-for-2-5-million-miles/.

16. Amory Lovins, *Reinventing Fire, Bold Business Solutions for the New Energy Era* (White River Junction, VT: Chelsea Green, 2011), 22–24.

17. John Fialka, "Advocacy: Entering a Climate-Changed World with Chutzpah and Bananas—Amory Lovins," *E&E News*, January 24, 2013, http://www.eenews.net/stories/1059975214. Amory Lovins, "Hypercars: Uncompromised Vehicles, Disruptive Technologies, and the Rapid Transition to Hydrogen" (slide presentation), presented to the California Air Resources Board, April 18, 2001, https://rmi.org/insight/hypercars-uncompromised-vehicles-disruptive-technologies-and-the-rapid-transition-to-hydrogen/.

18. Wikipedia, s.v. "Nissan Leaf," accessed August 27, 2015, https://en.wikipedia.org/wiki/Nissan_Leaf; Kelly Lin and Jade Nelson, "2019 Nissan Leaf Plus First Test: Quicker Than You Might Expect," *Motortrend*, April 25, 2019, https://www.motortrend.com/cars/nissan/leaf/2019/2019-nissan-leaf-plus-first-test-review/; "2019 Nissan Leaf (62 kW-hr battery pack)," U.S. Department Energy, Energy Efficiency and Renewable Energy, accessed May 22, 2020, https://www.fueleconomy.gov/feg/Find.do?action=sbs&id=41276.

19. Wikipedia, s.v. "Tesla, Inc.," accessed August 22, 2019, https://en.wikipedia.org/wiki/Tesla,_Inc..

20. Barrie Dickenson, "Cut from a Different Cloth," Tesla, February 15, 2007, https://www.tesla.com/blog/cut-different-cloth.

21. "About Tesla," Tesla Motors, accessed August 27, 2015, http://www.tesla-motors.com/about.

22. "2017 Chevrolet Volt EV," Fuelefficiency.com, Energy Efficiency and Renewable Energy, U.S. Department of Energy, accessed August 15, 2017, https://www.fueleconomy.gov/feg/Find.do?action=sbs&id=38187.

23. Alex Davis, "How GM Beat Tesla to the First True Mass-Market Electric Car," *Wired*, February 2016, https://www.wired.com/2016/01/gm-electric-car-chevy-bolt-mary-barra/. Jeff Zurschmeide, "2017 Chevrolet Bolt Is a Game-Changer," *Portland Tribune*, August 10, 2017, A13.

24. "Electric Cars 101: The Answers to All Your EV Questions," *Consumer Reports*, July 21, 2019, https://www.consumerreports.org/hybrids-evs/electric-cars-101-the-answers-to-all-your-ev-questions/.

25. "Plug Share," Plug Share, accessed August 20, 2019, https://www.plugshare.com/.

26. Damien Wilde, "Google Maps Now Shows the Real Time Availability of Nearby EV Charging Stations in US, UK," *9to5Google*, April 23, 2019, https://9to5google.com/2019/04/23/google-maps-ev-charging-stations-availability/.

27. Steven Loveday, "Monthly Plug-In EV Sales Scorecard: July 2019," *Inside EV*, August 2, 2019, https://insideevs.com/news/362819/ev-sales-scorecard-july-2019/.

28. "USA—Flash Report, Sales Volume, 2018," *Automotive Industry Portal Marketlines*, accessed August 20, 2019, https://www.marklines.com/en/statistics/flash_sales/salesfig_usa_2018.

29. Wikipedia, s.v. "List of Electric Cars Currently Available," accessed August 20, 2019, https://en.wikipedia.org/wiki/List_of_electric_cars_currently_available.

30. Mike Berners-Lee and Duncan Clark, "What's the Carbon Footprint of . . . a New Car," *Guardian*, September 23, 2010, https://www.theguardian.com/environment/green-living-blog/2010/sep/23/carbon-footprint-new-car.

31. Jeanne Lee, "What Is the Total Cost of Owning a Car," *Nerd Wallet*, September 15, 2016, https://www.nerdwallet.com/blog/loans/total-cost-owning-car/.

32. Maddie Shephard, "Eleven Surprising Working from Home Statistics," *Futura*, July 23, 2019, https://www.fundera.com/resources/working-from-home-statistics; Derek Thompson, "The Workforce Is About to Change Dramatically," *Atlantic*, August 6, 2020, https://www.theatlantic.com/ideas/archive/2020/08/just-small-shift-remote-work-could-change-everything/614980/.

33. "Waze Carpool," Waze, accessed November 10, 2019, https://www.waze.com/carpool.

34. Kathryn Gessner, "From Avis to Zip Car: What's Driving Change in How We Get Around," *Second Measure*, January 22, 2018, https://secondmeasure.com/datapoints/from-avis-to-zipcar/.

35. "Let Your Car Work for You," Turo, accessed November 10, 2019, https://turo.com/en-us/list-your-car.

36. Leanna Garfield, "10 Cities That Are Starting to Go Car-Free," *Business Insider*, August 16, 2016, http://www.businessinsider.com/cities-going-car-free-2016-8/#oslo-norway-will-implement-its-car-ban-by-2019-1. Kristen Hauser, "Better Market Street," *Futurism*, October 17, 2019, https://futurism.com/the-byte/urban-trend-banning-cars-city-streets.

37. Wikipedia, s.v. "Public Transport," accessed September 1, 2015, https://en.wikipedia.org/wiki/Public_transport.

38. Wikipedia, s.v. "New York City Subway," accessed September 1, 2015, https://en.wikipedia.org/wiki/New_York_City_Subway.

39. Wikipedia, s.v. "Jaime Lerner," accessed September 2, 2015, https://en.wikipedia.org/wiki/Jaime_Lerner.

40. Wikipedia, s.v. "Bus Rapid Transit," accessed September 2, 2015, https://en.wikipedia.org/wiki/Bus_rapid_transit.

41. "Global BRT Data," EMBARQ, accessed October 23, 2019, http://brtdata.org/.

42. "Portland Catches the Orange Line," *Portland Tribune*, September 2015, 48.

43. James Strickland, "Energy Efficiency of Different Modes of Transportation," (lecture notes) Stanford University, updated February 12, 2009, http://adl.stanford.edu/aa260/Lecture_Notes_files/transport_fuel_consumption.pdf.

44. Wikipedia, s.v. "High-Speed Rail," accessed September 2, 2015, https://en.wikipedia.org/wiki/High-speed_rail.

45. "No Amtrak Isn't About to Turn a Profit," Bumburg Opinion, November 21, 2019, https://www.bloomberg.com/opinion/articles/2019-11-21/amtrak-isn-t-about-to-turn-a-profit.

46. Max Fischer, "The Market for High Speed Trains Stays the Course," Credit Suisse, February 12, 2015, https://www.credit-suisse.com/about-us-news/en/articles/news-and-expertise/the-market-for-high-speed-trains-stays-the-course-201502.html.

47. John Murlis, "Guidance to Natural Capital Partners on the Treatment of Offsetting for Air Travel in the CarbonNeutral Protocol," CarbonNeutral, November 2019, https://carbonneutral.com/pdfs/Revised-Guidance-Aviation-Impacts-John-Murlis-December-2019.pdf.

48. "High Speed Rail History," International Union of Railways, accessed September 3, 2015, http://www.uic.org/High-Speed-History#t1981-2009-HSR-services-spreading-in-the-world.

49. Frank Tong, "Full Speed Ahead for China's High-Speed Rail Network in 2019 in Bid to Boost Slowing Economy," *South China Morning Post*, January 3, 2019, https://www.scmp.com/economy/china-economy/article/2180562/full-speed-ahead-chinas-high-speed-rail-network-2019-bid-boost.

50. "Florida's Brightline Lights the Way Towards Fast, Successful Trains," High Speed Rail Alliance, accessed October 25, 2020, https://www.hsrail.org/florida-high-speed-rail.

51. "FRA Delivers Major Breakthrough for US Highspeed Rail with Texas Safety Standards," High Speed Rail Alliance, September 22, 2020, https://www.hsrail.org/fra-delivers-major-breakthrough-us-high-speed-rail-texas-safety-standards.

52. "About California High Speed Rail Authority," California High Speed Rail Authority, accessed September 4, 2015, http://www.hsr.ca.gov/About/index.html.

53. California Proposition 1A, High-Speed Rail Act (2008)," Ballotpedia, accessed October 26, 2020, https://ballotpedia.org/California_Proposition_1A,_High-Speed_Rail_Act_(2008); Ralph Vartabedian, "California High Speed Rail Faces Mounting Woes," *Los Angeles Times*, September 8, 2020, https://www.govtech.com/fs/transportation/California-High-Speed-Rail-Faces-Mounting-Financial-Woes.html.

54. "Summary of Travel Trends: 2017 National Household Travel Survey," Federal Highway Administration, U.S. Department of Transportation, July 2018, https://nhts.ornl.gov/assets/2017_nhts_summary_travel_trends.pdf.

55. "Mobility Services Becoming More Popular as Alternatives to Vehicle Ownership, According to COX Automotive Study," Cox Automotive, August 23, 2018, https://www.coxautoinc.com/news/evolution-of-mobility-study-alternatives-to-ownership/.

56. Benjamin Davis, Tony Dutzik, and Phineas Baxandall, "Transportation and the New Generation," Frontier Group, U.S. PIRG, April 2012, https://uspirg.org/sites/pirg/files/reports/Transportation%20&%20the%20New%20Generation%20vUS_0.pdf; Christopher Knittel and Elizabeth Murphy, "Generational Trends in Vehicle Ownership and Use: Are Millennials Any Different?" Massachusetts Institute of Technology Center for Energy and Environmental Policy

Research, CEEPR WP 2019-006, February 8, 2018, http://ceepr.mit.edu/files/papers/2019-006.pdf.

57. "New Report from Allison+Partners Uncovers a Shift from Car Culture to Mobility Culture," (press release), Allison+Partners, March 13, 2019, https://s3.us-west-2.amazonaws.com/allisonpr.com/201903/0/f/6/20190311155508_9094345/MobilityIPRelease.pdf. This survey was of 16-to-24-year-olds, taken in January 2018.

58. Lee Miller and Wei Lu, "Gen Z Is Set to Outnumber Millennials within a Year," *Bloomberg*, August 20, 2018, https://www.bloomberg.com/news/articles/2018-08-20/gen-z-to-outnumber-millennials-within-a-year-demographic-trends.

59. Michael Manville, Brian D. Taylor, and Evelyn Blumenberg, "Falling Transit Ridership: California and Southern California," UCLA Institute of Transportation Studies, January 2018, https://www.its.ucla.edu/2018/01/31/new-report-its-scholars-on-the-cause-of-californias-falling-transit-ridership/.

60. Home page, RideAmigos, accessed November 11, 2019, https://rideamigos.com/.

CHAPTER 14

1. "Areas of Work," Northwest Energy Efficiency Alliance, accessed April 27, 2020, https://neea.org/our-work.

2. "About IMT," Institute of Market Transformation, accessed March 5, 2020, https://www.imt.org/about/.

3. "City Energy: A Joint Project of NRDC + IMT," (fact sheet), Institute of Market Transformation, accessed March 5, 2020, https://www.cityenergyproject.org/wp-content/uploads/2019/01/About_City_Energy_Project_Fact_Sheet.pdf.

4. "Windpower Capacity Worldwide Reaches 597 GW, 50.1 Added in 2018," (press release), World Wind Energy Association, February 25, 2019, https://wwindea.org/blog/2019/02/25/wind-power-capacity-worldwide-reaches-600-gw-539-gw-added-in-2018/. "World Market Has Reached 486 Gigawatts from Where 54 GW Has Been Installed Last Year," World Wind Energy Association, June 8, 2017, http://www.wwindea.org/11961-2/.

5. "Global Wind Industry Fact Sheet 2020: Top 10 Largest Wind Turbine Manufacturers," *BizVibe*, May 26, 2020, https://www.bizvibe.com/blog/energy-and-fuels/top-10-wind-turbine-manufacturers-world/.

6. "State Renewable Portfolio Standards and Goals," National Conference of State Legislatures, December 31, 2019, https://www.ncsl.org/research/energy/renewable-portfolio-standards.aspx.

7. "New Jersey's Clean Energy Picture," New Jersey Department of Environmental Protection, updated February 18, 2020, https://www.nj.gov/dep/aqes/opea-clean-energy.html.

8. R. L. Hills, *Power from Wind: A History of Windmill Technology* (Cambridge: Cambridge University Press, 1996). (This technology section presents a summarized account that was presented in book-length form.)

9. "Wind Facts at a Glance," American Wind Energy Association, accessed May 22, 2020, https://www.awea.org/wind-101/basics-of-wind-energy/wind-facts-at-a-glance.

10. Elizabeth Weise and Rick Jervis, "As Climate Threat Looms, Texas Republicans Have a Solution: Giant Windfarms Everywhere," *USA Today*, October 18, 2019, https://www.usatoday.com/story/news/2019/10/18/texas-wind-energy-so-strong-its-beating-out-coal-power/3865995002/.

11. Taste of Travel, "Mappandal Wind Farm in Tamilnadu," YouTube video, 9:22, December 3, 2019, https://www.youtube.com/watch?v=2YZB4ArAOy4.

12. "Top Ten Biggest Wind Farms," *Power Technology*, June 29, 2019, https://www.power-technology.com/features/feature-biggest-wind-farms-in-the-world-texas/.

13. Wikipedia, s.v. "List of Offshore Wind Farms," last modified October 23, 2020, http://en.wikipedia.org/wiki/List_of_offshore_wind_farms.

14. John Parnell, "World's Largest Floating Wind Turbine Begins Generating Power," *gtm* (Greentech Media), https://www.greentechmedia.com/articles/read/worlds-largest-floating-wind-turbine-connected.

15. "Germany 2020, Energy Policy Review," International Energy Agency, February 2020, https://www.iea.org/reports/germany-2020.

16. "Best Research Cell Efficiencies," (chart) National Renewable Energy Laboratory, accessed October 25, 2020, https://www.nrel.gov/pv/cell-efficiency.html.

17. "Solar Energy Research Data," Solar Energy Industries Association, accessed March 2, 2020, https://www.seia.org/solar-industry-research-data.

18. Brian McConnell, "Solar Energy: This Is What a Disruptive Technology Looks Like," *Resilience*, April 25, 2013, https://www.resilience.org/stories/2013-04-25/solar-energy-this-is-what-a-disruptive-technology-looks-like/.

19. Mathias Aarre Maehium, "What's the Difference Between Net Metering and Feed-in Tariffs?" *Energy Informative*, March 15, 2014, http://energyinformative.org/net-metering-feed-in-tariffs-difference.

20. "United States Surpasses 2 Million Solar Installations," (press release), Solar Energy Industries Association, May 9, 2019, https://www.seia.org/news/united-states-surpasses-2-million-solar-installations.

21. Josh Sanburn, "A Burst of Energy, Inside the World's Largest Solar Power Plant," *Time*, March 9, 2015; "Just the Facts: Topaz Solar Farms," BHE Renewables, February 2020, https://www.bherenewables.com/include/pdf/fact_sheet_topaz.pdf.

22. "What Is U.S. Electric Generation by Source?" U.S. Energy Information Agency, updated February 27, 2020, https://www.eia.gov/tools/faqs/faq.php?id=427&t=3.

23. Gaëtan Masson, ed., "Snapshot of the Global PV Markets 2020, International Energy Agency," IEA-PVPS T1-37:2020, April 2020, https://iea-pvps.org/wp-content/uploads/2020/04/IEA_PVPS_Snapshot_2020.pdf.

24. Angus McCrone, "Energy, Vehicles, Sustainability—10 Predictions for 2020," *BloombergNEF*, January 16, 2020, https://about.bnef.com/blog/energy-vehicles-sustainability-10-predictions-for-2020/.

25. Marlene Motyka, Carolyn Amon, and Andrew Slaughter, "Global Renewable Energy Trends: Solar and Wind Move from Mainstream to Preferred," *Deloitte Insights*, September 13, 2018, https://www2.deloitte.com/us/en/insights/industry/power-and-utilities/global-renewable-energy-trends.html.

26. "Renewable Energy Now Accounts for a Third of the Global Installed Capacity," (press release), International Renewable Energy Agency, April 2, 2019, https://www.irena.org/newsroom/pressreleases/2019/Apr/Renewable-Energy-Now-Accounts-for-a-Third-of-Global-Power-Capacity.

27. Amory Lovins, *Small Is Profitable: The Hidden Economic Benefits of Making Electrical Resources the Right Size* (Basalt, CO: Rocky Mountain Institute, 2002).

28. *BPA Distributed Energy Resource Benchmarking Report* (Portland, OR: Bonneville Power Administration, December 2017), https://www.bpa.gov/EE/Technology/demand-response/Documents/2017_Distributed_Energy_Resources_Benchmarking_Report.pdf.

29. Roberto Rodriguez, "Distributed Energy Resources Set to Flourish in Europe," (reported in) *ROC Nederland*, December 18, 2017, http://www.topofthejobs.org/default.php?fr=nieuws&nieuwsitem=102731.

30. Garrett Fitzgerald, Chris Nelder, and James Newcomb, "Electric Vehicles as Distributed Energy Resources," Electricity Innovation Lab, Rocky Mountain Institute, accessed September 23, 2018, https://rmi.org/wp-content/uploads/2017/04/RMI_Electric_Vehicles_as_DERs_Final_V2.pdf.

31. Dean Sigler, "SAS 2019: Larry Cooke and NovaSolix," *Sustainable Skies*, November 16, 2019, http://sustainableskies.org/sas-2019-larry-cooke-novasolix/.

32. "PacWav," U.S. Office of Energy Efficiency and Renewable Resources, Water Power Technologies Office, accessed March 4, 2020, https://www.energy.gov/eere/water/pacwave.

33. "U.S. Fusion Research Sites," US ITER Project Office, U.S. Department of Energy, accessed March 4, 2020, https://www.usiter.org/fusion/us-fusion-research-sites.

CHAPTER 15

1. Julie Steward Williams, *From the Mountains to the Sea, Early Hawaiian Life* (Honolulu, HI: Kamehameha Schools Press, 1997), 9–29, accessed via *Ulukau: The Hawaiian Electronic Library*, http://ulukau.org/elib/cgi-bin/library?c=english&l=en.

2. "Ahupua'a," HawaiiHistory.org, accessed October 15, 2015, http://www. hawaiihistory.org/index.cfm?fuseaction=ig.page&CategoryID=299#top.

3. Douglas Brinkley, *The Quiet World: Saving Alaska's Wilderness Kingdom 1879–1960* (New York: HarperCollins, 2011), 491.

4. David Bacon and David B. Williams, "Restraint and Hope: Lessons for the Lake Baikal and Arctic National Wildlife Refuge," Truthout, January 11, 2011, http://truth-out.org/archive/component/k2/item/93890:restraint-and-hope-les sons-from-lake-baikal-and-the-arctic-national-wildlife-refuge.

5. "Potential for Oil Production from the Coastal Plain of the Arctic National Wildlife Refuge," Energy Information Administration, accessed June 12, 2016, http://www.atsummit50.org/session/plenary1-3.html.

6. "Proposal to Drill in the Arctic National Wildlife Refuge," *Congressional Almanac*, 1991, http://library.cqpress.com/cqalmanac/document. php?id=cqal91-1110326; Michael Kraft and Diana Wuertz, "Environmental Advocacy in the Corridors of Government," in *The Symbiotic Earth: Discourse and Our Creation of the Environment*, ed. James G. Cantrill and Christine L. Oravei (Lexington: University Press of Kentucky, 1996), 109–10.

7. M. Alex Johnson, "Greenpeace Protesters Blocking Oil Ship Rappel Down from Portland Bridge," *NBC News*, July 30, 2015, http://www.nbcnews.com/ news/us-news/greenpeace-protesters-blocking-oil-ship-rappel-down-portland-bridge-n401421.

8. Juliet Eilperin and Steven Mufson, "Royal Dutch Shell Suspends Arctic Drilling Indefinitely," *Washington Post*, September 28, 2015, http://www.wash ingtonpost.com/news/energy-environment/wp/2015/09/28/royal-dutch-shell-suspends-arctic-drilling-indefinitely/.

9. Wikipedia, s.v. "CITES," accessed April, 18, 2015, http://en.wikipedia. org/wiki/CITES.

10. "In 2003, 30 Years of International Agreement," *CITES World*, March 3, 2003, http://www.cites.org/sites/default/files/eng/news/world/30special.pdf.

11. "Alligator Snapping Turtle," *National Geographic*, accessed April 16, 2016, http://animals.nationalgeographic.com/animals/reptiles/alligator-snapping-turtle/.

12. *Convention on International Trade in Endangered Species of Wild Fauna and Flora* (text of treaty), Convention on International Trade in Endangered Species of Wild Fauna and Flora, signed 1973, amended 1979 and 1983, http://www.cites.org/ eng/disc/text.php.

13. Jan Kueera, "Czech Republic," *CITES World*, Issue 15, July 2005, https:// www.cites.org/sites/default/files/eng/news/world/15.pdf.

14. Yuan Jiming, "China," *CITES World*, Issue 15, July 2005, https://www. cites.org/sites/default/files/eng/news/world/15.pdf.

15. "Lacey Act," U.S. Fish and Wildlife Service, accessed March 11, 2020, https://www.fws.gov/international/laws-treaties-agreements/us-conservation-laws/lacey-act.html. Sandra Cleva, "United States of America—The Lacey Act,"

CITES World, Issue 15, July 2005, https://www.cites.org/sites/default/files/eng/news/world/15.pdf.

16. Tim Van Norman, "United States of America—The Endangered Species Act," *CITES World*, Issue 15, July 2005, https://www.cites.org/sites/default/files/eng/news/world/15.

17. William Dietrich, *The Final Forest: The Battle for the Last Great Trees of the Pacific Northwest* (New York: Penguin, 1992), 98.

18. "Permaculture Neighborhood Center," Permaculture Research Institute, updated July 10, 2015, https://permacultureglobal.org/projects/480-permaculture-neighborhood-center.

19. "Worldwide Permaculture Projects," Permaculture Worldwide Network, accessed November 22, 2019, https://permacultureglobal.org/projects?page=1.

20. "The PRI Restarts the Permaculture Teacher Registry," Permaculture Research Institute, October 19, 2011, http://permaculturenews.org/2011/10/19/the-pri-restarts-the-permaculture-teacher-registry/.

21. "Timeline," Food Not Lawns, accessed June 20, 2016, http://www.foodnotlawns.com/about-us.html.

22. "Directory of Local Chapters," Food Not Lawns, accessed June 20, 2016, http://www.foodnotlawns.com/local-chapters.html.

23. "Certification Criteria," Audubon Society of Portland (Oregon), accessed August 8, 2014, http://audubonportland.org/issues/backyardhabitat/criteria.

24. "Certify to Show Your Commitment to Wildlife," National Wildlife Federation, accessed January 12, 2020, https://www.nwf.org/Garden-for-Wildlife/Certify.

25. Rebecca Philip, "Bioremediation: The Pollution Solution?" Microbiology Society, December 8, 2015, https://microbiologysociety.org/blog/bioremediation-the-pollution-solution.html.

26. "Environmental Remediation Using Mycelium," YouTube video, 2:23, Paul Staments, June 10, 2013, https://www.youtube.com/watch?v=lJHXkfNCl5E&noredirect=1.

27. Renee Alexander, "Oyster Mushrooms Helped Clean Up after California's Wildfires. Why Is It So Hard to Make a Business Case for Mycoremediation," *New Food Economy*, February 18, 2019, https://newfoodeconomy.org/mycoremediation-radical-mycology-mushroom-natural-disaster-pollution-clean-up/.

28. "Mycoremediation, Division of Environmental Enhancement (DEE)," Mushroom Mountain, accessed September 30, 2019, https://mushroommountain.com/mycoremediation/. Maddie Stone, "The Plan to Clean Up the World's Largest Oil Spill with Fungus," *Vice*, March 5, 2015, https://www.vice.com/en_us/article/jp5k9x/the-plan-to-mop-up-the-worlds-largest-oil-spill-with-fungus.

29. Jaclyn Saorsail, "Mycoremediation Project," CommonWealth Urban Farms, accessed September 29, 2019, http://commonwealthurbanfarms.com/mycoremediation-project/.

30. Andrew Stevenson, "Scientists Find Fungus with an Appetite for Plastic in Rubbish Dump," *Agroforestry World*, accessed October 1, 2019, https://blog.worldagroforestry.org/index.php/2017/09/12/scientists-find-fungus-appetite-plastic-rubbish-dump/.

31. "Online Mycology Courses: Teaching the Language of Fungi," Mycologos, accessed October 26, 2020, https://mycologos.world/.

32. "Bioremediation Companies," Environmental Expert, accessed July 27, 2015, http://www.environmental-expert.com/soil-groundwater/bioremediation/companies/page-1.

33. "Scientists," Waterstone Environmental, accessed January 11, 2020, http://www.waterstone-env.com/scientists.

34. "Bioremediation: Nature's Way to a Cleaner Environment," United States Geological Survey, accessed July 31, 2014, http://water.usgs.gov/wid/html/bioremed.html.

35. "CERCLA Overview," U.S. Environmental Protection Agency, accessed August 21, 2014, http://www.epa.gov/superfund/policy/cercla.htm.

36. "Superfund History," Environmental Protection Agency, accessed October 26, 2020, https://www.epa.gov/superfund/superfund-history.

37. "National Oil and Hazardous Substances Contingency Plan (NCP)," U.S. Environmental Protection Agency, accessed August 9, 2015, http://www2.epa.gov/emergency-response/national-oil-and-hazardous-substances-pollution-contingency-plan-ncp-overview.

38. "Superfund Redevelopment Initiative," U.S. Environmental Protection Agency, accessed August 6, 2015, http://www.epa.gov/superfund/programs/recycle/index.html.

39. "Reuse Examples—Green Space," U.S. Environmental Protection Agency, accessed August 7, 2015, http://www.epa.gov/superfund/programs/recycle/live/greenspace.html.

40. Jeff McMenemy, "Portsmouth's Amico Gets EPA Honor for PFAS Fight," Seacoastonline, updated November 8, 2018, https://www.seacoastonline.com/news/20181107/portsmouths-amico-gets-epa-honor-for-pfas-fight.

41. Cassie Cohen and Donovan Smith, "My View: Pressure Led to Better Cleanup Plan," *Portland Tribune*, January 26, 2017, https://pamplinmedia.com/pt/10-opinion/341738-221739-my-view-public-pressure-led-to-better-harbor-plan.

42. "Superfund National Priorities List (NPL)," U.S. Environmental Protection Agency, updated November 8, 2019, https://www.epa.gov/superfund/superfund-national-priorities-list-npl; "Superfund: Transforming Communities, Accomplishments Report—FY2018," U.S. Environmental Protection Agency, accessed October 14, 2020, https://semspub.epa.gov/work/HQ/100001884.pdf.

43. "Our Story," African Parks, accessed October 26, 2020, https://www.africanparks.org/about-us/our-story.

44. "15 Ways to Help Protect Endangered Species," Endangered Species Coalition, accessed March 11, 2020, https://www.endangered.org/15-ways-to-help-protect-endangered-species/.

45. "Certify Your Commitment to Wildlife," National Wildlife Federation, accessed January 12, 2020, https://www.nwf.org/Garden-for-Wildlife/Certify.

46. "iNaturalist," California Academy of Sciences and *National Geographic*, accessed October 1, 2019, https://www.inaturalist.org/.

47. "Anatomy of Brownfields Redevelopment," U.S. Environmental Protection Administration, EPA-560-06-245, October 2006, https://www.epa.gov/sites/production/files/2015-09/documents/anat_bf_redev_101106.pdf.

CHAPTER 16

1. "2013 Annual Report," Johnson Creek Watershed Council, accessed August 21, 2014, http://jcwc.org/wp-content/uploads/2012/01/annual-report-2013_fnl-v3.pdf.

2. "The Clearwater Story," Clearwater, accessed August 21, 2014, http://www.clearwater.org/about/the-clearwater-story/.

3. "How Is the Hudson Doing?" New York Department of Environmental Conservation, accessed August 21, 2014, http://www.dec.ny.gov/lands/77105.html.

4. "Mid-Hudson Regional Sustainability Plan," Orange County, New York, accessed October 27, 2020, https://www.orangecountygov.com/300/Mid-Hudson-Regional-Sustainability-Plan.

5. "Hudson River PCBs," Clearwater, accessed August 21, 2014, http://www.clearwater.org/ea/pcb-contamination/.

6. "Annual Report 2018," Waterkeeper Alliance, accessed October 14, 2020, https://waterkeeper.org/wp-content/uploads/2018/12/Waterkeeper-Alliance-FY18-Annual-Report.pdf.

7. "Spectacular Time Lapse Dam 'Removal'" YouTube video, *National Geographic*, 2:02, November 2, 2011, https://www.youtube.com/watch?v=4LxMHmw3Z-U.

8. "Return of the River" (trailer for *Return of the River* movie), 2:59, accessed January 12, 2020, http://elwhafilm.com/. This relates to removing two dams on the Elwha River.

9. Amy Souers Kober, "Twenty Years of Dam Removal Successes—and What's Up Next," American Rivers, June 27, 2019, https://www.americanrivers.org/2019/06/twenty-years-of-dam-removal-successes-and-whats-up-next/.

10. Tara Lohan, "The Removal of One Maine Dam 20 Years Ago Changed Everything," *Earth Island Journal*, February 12, 2019, https://www.earthisland.org/journal/index.php/articles/entry/removal-maine-dam-river-conservation.

11. Jacques Leslie, "Oregon's Klamath River Basin One Step Closer to Historic Dam Removal," *Earth Island Journal*, April 17, 2014, http://www.earthisland.org/journal/index.php/elist/eListRead/oregons_klamath_river_basin_one_step_closer_to_historic_dam_removal/.

12. "Water Restoration Certificates—U.S.," BEF, accessed July 24, 2019, https://store.shrinkyourfoot.org/products/water-restoration-certificates-blend.

13. "Water Restoration Certificates—U.S."

14. Sandra Postal, "Once a Smelly Nuisance, Mexicali's Wastewater Now Brings Life to the Colorado Delta," *National Geographic*, May 9, 2013, http://voices.nationalgeographic.com/2013/05/09/once-a-smelly-nuisance-mexicalis-wastewater-now-brings-life-to-the-colorado-delta/; "Water," Silk, WhiteWave Services, accessed December 8, 2015, https://silk.com/yaywater.

15. "Jesse Creek, ID," Bonneville Environmental Foundation, accessed December 9, 2015, http://www.b-e-f.org/project-portfolio/jesse-creek/.

16. "Roaring Fork River," Bonneville Environmental Foundation, accessed December 9, 2015, http://www.b-e-f.org/project-portfolio/roaring-fork-river/.

17. "Intel Corporation Restore Water Goal 2018 Annual Report," LimnoTech, prepared for Intel, April 2019, https://www.intel.com/content/www/us/en/environment/restore-water-goal-report.html.

18. "BEF Water Restoration Program," Bonneville Environmental Foundation, accessed August 14, 2018, http://www.b-e-f.org/environmental-products/water-restoration-certificates/.

19. "Frequently Asked Questions about the Change the Course Campaign," Bonneville Environmental Foundation, accessed December 9, 2015, http://www.b-e-f.org/faq/.

20. "How Many Times Did the Cuyahoga River Burn," (video), Cuyahoga Valley National Park, National Park Service, accessed February 26, 2016, http://www.nps.gov/cuva/upload/7b_3.swf.

21. James McCarty, "Cuyahoga River Cleanup Reaches New Benchmark with Walleye Discovery," *Plain Dealer*, September 13, 2015, http://www.cleveland.com/metro/index.ssf/2015/09/cuyahoga_river_water_improveme.html.

22. "The Cuyahoga River," Cuyahoga Valley National Park, National Park Service, accessed February 26, 2016, http://www.nps.gov/cuva/learn/kidsyouth/the-cuyahoga-river.htm.

23. "2015 Ohio Sport Fish Health and Consumption Advisory," Ohio Environmental Protection Agency, accessed February 26, 2016, http://www.epa.state.oh.us/dsw/fishadvisory/index.aspx#145214734-statewide.

24. "About Us," Friends of Los Angeles River, accessed October 26, 2020, https://folar.org/about-us/#history; Hillary Rosner, "Los Angeles River: From Concrete Ditch to Urban Oasis," *National Geographic*, July 18, 2014, http://news.nationalgeographic.com/news/2014/07/140719-los-angeles-river-restoration-kayaking-greenway/; "LA River Ecosystem Restoration," Los Angeles River

Vitalization, City of Los Angeles, accessed October 26, 2020, http://lariver.org/blog/la-river-ecosystem-restoration.

25. "Muslims Spearhead Environmental Efforts to Clean Up Cooks River in Sydney Australia," *Muslim Environment Watch*, October 28, 2010, https://muslimenvironment.wordpress.com/category/australia/muslims-spearhead-environmental-efforts-to-clean-up-cooks-river-in-sydney-australia/.

26. Alistair Driver, "The Thames—Recovery from Biological Death," (slide show), *Environmental Agency*, England, accessed August 21, 2014, http://www.restorerivers.eu/Portals/27/Alastair%20Driver.pdf.

27. Ruth Greenspan Bell and Libor Jansky, "Public Participation in the Danube Management of the Danube River: Necessary but Neglected," in *Public Participation in the Governance of Freshwater Resources*, ed. Carl Brush et al. (Tokyo: United Nations University Press, 2005), 108–11.

28. "The Green Danube Partnership," Coca-Cola Hellenic Bottling, accessed December 10, 2015, http://www.coca-colahellenic.com/sustainability/community/waterandenvironment/greendanubepartner.

29. Rebecca Staudenmaier, "Europewide Rhine River Cleanup Draws Thousands of Volunteers," *Deutsche Welle*, September 15, 2018, https://www.dw.com/en/europewide-rhine-river-cleanup-draws-thousands-of-volunteers/a-45499888. "Father Rhine's Recovery," What's Up Germany, accessed October 26, 2020, https://whatsupgermany.de/father-rhines-recovery/.

30. Home page, Ganga Action Parivar, accessed January 13, 2020, http://www.gangaaction.org/hych/.

CHAPTER 17

1. Oregon Encyclopedia, s.v., "Oregon Beach Bill," updated August 26, 2020, https://www.oregonencyclopedia.org/articles/oregon_beach_bill.

2. Wikipedia, s.v., "Texas Open Beaches Act," accessed August 27, 2014, http://en.wikipedia.org/wiki/Texas_Open_Beaches_Act.

3. "Public Access Rights," Sea Grant, University of Hawaii, accessed December 9, 2016, http://seagrant.soest.hawaii.edu/public-access-rights.

4. "Trash Free Seas Alliance," Ocean Conservancy, accessed August 23, 2014, http://www.oceanconservancy.org/our-work/trash-free-seas-alliance/.

5. "Building a Clean Swell: 2018 Annual Report," Ocean Conservancy, accessed August 9, 2019, https://oceanconservancy.org/wp-content/uploads/2018/07/Building-A-Clean-Swell.pdf. "International Coastal Cleanup," Ocean Conservancy, accessed August 27, 2014, http://www.oceanconservancy.org/our-work/international-coastal-cleanup/.

6. Tick Root, "What's the World's Most Littered Plastic Pollutant? Cigarette Butts," *National Geographic*, August 9, 2019, https://www.nationalgeographic.com/environment/2019/08/cigarettes-story-of-plastic/.

7. "California Marine Protected Areas," California Department of Fish and Wildlife, October 26, 2020, https://wildlife.ca.gov/Conservation/Marine/MPAs 1.

8. "National Marine Sanctuary System," National Oceanic and Atmospheric Administration, accessed August 24, 2014, http://sanctuaries.noaa.gov/.

9. Cynthia Barnett, "Saving the Seas," *National Geographic*, February 2017, 54–75.

10. "It's a Keeper, the Law That's Saving American Fisheries," (booklet), Ocean Conservancy, January 9, 2014, http://issuu.com/jyeary/docs/ff-msa-report-2013-final?e=10446736/6304439#search.

11. "It's a Keeper," 14–15.

12. S. A. Murawski, "Rebuilding Depleted Fish Stocks: The Good, the Bad, and Mostly, the Ugly," *ICES Journal of Marine Science* 67 (2010), 1830–40, in "It's a Keeper, the Law That's Saving American Fisheries," (booklet), The Ocean Conservancy, January 9, 2014, http://issuu.com/jyeary/docs/ff-msa-report-2013-final? e=10446736/6304439#search.

13. Emma L. Hickerson, George Schmahl, Kathy Broughton, and Steven Gittings, *Flower Garden Banks National Marine Sanctuary Condition Report 2008* (Silver Springs, MD: U.S. Department of Commerce, National Oceanic and Atmospheric Administration, Office of Marine Sanctuaries), September 2008, https://nmsflow ergarden.blob.core.windows.net/flowergarden-prod/media/archive/document_li brary/scidocs/fgbnmsconditionreport.pdf.

14. Robin Kundis Craig, "The Gulf Oil Spill and National Marine Sanctuaries," *Environmental Law Reporter* 40, ELR 11078, November 2010.

15. "Status of and Threat to Coral Reefs," International Coral Reef Initiative, accessed October 26, 2020, https://www.icriforum.org/about-coral-reefs/status-of-and-threat-to-coral-reefs/.

16. Tony Juniper, *What Has Nature Ever Done for Us? How Money Really Does Grow on Trees* (Santa Fe, NM: Energetic Press, 2013), 192–93; Paige Henry and Sarah Crandall, "Sea Otters Can Help Reduce Carbon Dioxide in the Atmosphere," *Project Censored*, January 29, 2013, http://www.projectcensored.org/sea-otters-can-help-reduce-carbon-dioxide-in-the-atmosphere/.

17. "Sea Otter Conservation," SeaOtters.com, accessed October 26, 2020, https://www.seaotters.com/sea-otter-conservation/; Michael McPhate, "California Today: The Plight of the Sea Otter," *New York Times*, October 26, 2017, https://www.nytimes.com/2017/10/26/us/california-today-the-plight-of-the-sea-otter.html.

18. "History and Purpose," International Whaling Commission, accessed September 3, 2014, http://iwc.int/history-and-purpose.

19. "Status of Whales," International Whaling Commission, accessed September 3, 2014, http://iwc.int/status.

20. "Marine Mammal Protection Act (MMPA)," NOAA Fisheries, National Oceanic and Atmospheric Administration, accessed October 26, 2020, https://

www.fisheries.noaa.gov/national/marine-mammal-protection/marine-mammal-protection-act.

21. Wikipedia, s.v. "Endangered Species Act," accessed August 30, 2014, http://en.wikipedia.org/wiki/Endangered_Species_Act.

22. Aubrey Wallace, *Eco-Heroes, Twelve Tales of Environmental Victory* (San Francisco, CA: Mercury House, 1993), 37–58.

23. "About STC: Organizational Background," Sea Turtle Conservancy, accessed October 28, 2020, https://conserveturtles.org/about-stc-organizational-background/.

24. "Tortuguero Sea Turtles," Costa Rica Exotica Natural, accessed August 9, 2019, https://www.tortugueroinfo.com/usa/sea_turtles_tortuguero.htm.

25. David Godfrey, "STC Programs: Research Programs: Tortuguero Program Is a Conservation Success," (video of presentation to Smithsonian Institution), Turtle Conservancy, May 20, 2009, https://conserveturtles.org/stc-programs-research-tortuguero-program-conservation-success/.

26. "International Convention for the Prevention of Pollution from Ships," International Maritime Organization, accessed September 12, 2014, http://www.imo.org/About/Conventions/ListOfConventions/Pages/International-Convention-for-the-Prevention-of-Pollution-from-Ships-(MARPOL).aspx.

27. "Marine Debris," National Oceanic and Atmospheric Administration, accessed August 22, 2014, http://marinedebris.noaa.gov/about-our-program/marine-debris-act; "Marine Debris Reauthorization Act," Surfrider Foundation, accessed August 22, 2014, http://www.surfrider.org/campaigns/entry/marine-debris-reauthorization-act.

28. "Princess Cruise Lines and Its Parent Company Plead Guilty to Environmental Probation Violations, Ordered to Pay $20 Million Criminal Penalty," *Justice News*, U.S. Department of Justice, June 3, 2019, https://www.justice.gov/opa/pr/princess-cruise-lines-and-its-parent-company-plead-guilty-environmental-probation-violations.

29. "The Helsinki Convention," Helcom, accessed October 28, 2020, https://helcom.fi/about-us/convention/; "HELCOM (Helsinki Commission) and Helsinki Convention," Marine Species, last edited September 2, 2020, http://www.marinespecies.org/introduced/wiki/HELCOM_(Helsinki_Commission)_and_Helsinki_Convention.

30. "Register of International Treaties and Other Agreements in the Field of the Environment," United Nations Environment Programme, December 30, 2005, https://www.unenvironment.org/resources/report/register-international-treaties-and-other-agreements-field-environment.

31. Jenna R. Jambeck, Roland Geyer, Chris Wilcox, Theodore R. Siegler, Miriam Perryman, Anthony Andrady, Ramani Narayan, and Kara Lavender Law, "Plastic Waste Inputs from Land into the Ocean," *Science* 347, no. 6223 (2015): 768–71, https://www.iswa.org/fileadmin/user_upload/Calendar_2011_03_AMERICANA/Science-2015-Jambeck-768-71__2_.pdf.

32. "State Plastic and Paper Bag Legislation," National Conference of State Legislatures, accessed January 13, 2020, https://www.ncsl.org/research/environment-and-natural-resources/plastic-bag-legislation.aspx.

33. "State Plastic and Paper Bag Legislation."

34. Joseph Kiprop, "Which Countries Have Banned Plastic Bags," *WorldAtlas*, accessed August 29, 2018, https://www.worldatlas.com/articles/which-countries-have-banned-plastic-bags.html. Caroll Excell, "127 Countries Now Regulate Plastic Bags. Why Aren't We Seeing Less Pollution?" World Resources Institute, March 11, 2019, https://www.wri.org/blog/2019/03/127-countries-now-regulate-plastic-bags-why-arent-we-seeing-less-pollution.

35. Neha Dasgupta and Mayank Bhardwaj, "Exclusive: India Set to Outlaw Six Single-Use Plastic Products on October 2—Sources," *Reuters*, August 2, 2019, https://www.reuters.com/article/us-india-pollution-plastic-exclusive/exclusive-india-set-to-outlaw-six-single-use-plastic-products-on-october-2-sources-idUSKCN1VI19F.

36. Jacob Silverman, "Why Is the World's Biggest Landfill in the Pacific Ocean?" *How Stuff Works*, accessed September 12, 2014, http://science.howstuffworks.com/environmental/earth/oceanography/great-pacific-garbage-patch.htm.

37. Sarah Griffiths, "Could a Teenager Save the World's Ocean? Student, 19, Claims His Invention Could Clean Up the Seas in Just Five Years," *London Daily Mail*, September 9, 2013, http://www.dailymail.co.uk/sciencetech/article-2415889/Boyan-Slat-19-claims-invention-clean-worlds-oceans-just-years.html.

38. "Crowdfunding Campaign to Ocean Cleanup Successfully Completed," Ocean Cleanup, accessed October 28, 2020, https://theoceancleanup.com/press/press-releases/crowd-funding-campaign-the-ocean-cleanup-successfully-completed/.

39. "End of Mission One—Plastic on Shore," (video), Ocean Cleanup, December 12, 2019, https://theoceancleanup.com/updates/.

40. Gary Gagliardi, "How Long Did It Take for Amazon to Break Even in the US?" Quora, September 7, 2017, https://www.quora.com/How-long-did-it-take-for-Amazon-to-break-even-in-the-US.

41. Emily Cardiff, "5 Fantastic Organizations Fighting to Protect Endangered Species," *One Green Planet*, May 16, 2014, http://www.onegreenplanet.org/animalsandnature/fantastic-organizations-fighting-to-protect-endangered-species/; "Manatee Facts," Save the Manatee Club, accessed October 28, 2017, http://www.savethemanatee.org/manfcts.htm.

42. "Citizen Science at NOAA," National Oceanic and Atmospheric Administration, U.S. Department of Commerce, accessed January 13, 2019, https://oceanservice.noaa.gov/news/apr15/volunteer.html.

43. Sarah Goddard, "How to Pass a Plastic Bag Ban: 8 Key Lessons," *Green That Life*, March 17, 2019, https://greenthatlife.com/plastic-bag-ban/.

44. Root, "Most Littered Plastic Pollutant?"

45. "Marine Biology Degree Programs in the U.S. (by State)," MarineBio Conservation Society, updated January 25, 2019, https://marinebio.org/careers/us-schools/.

CHAPTER 18

1. "Impact," Goodwill Industries International, accessed July 29, 2019, https://www.goodwill.org/annual-report/.

2. "Resale Industry and Statistics," NARTS: The Professional Association of Resale Professionals, accessed January 13, 2016 ($13.5 billion in annual revenues) and July 30, 2019 ($17.5 billion), http://www.narts.org/i4a/pages/index.cfm?pageid=3285.

3. "Beyond Waste Success Stories," Recology, accessed June 22, 2020, https://www.recology.com/success-stories/#success-stories.

4. "National Overview: Facts and Figures on Materials, Wastes and Recycling," Environmental Protection Agency, accessed October 28, 2020, https://www.epa.gov/facts-and-figures-about-materials-waste-and-recycling/national-overview-facts-and-figures-materials#NationalPicture.

5. "The Paper Recycling Success Story," Paper Recycles, accessed April 11, 2020, https://www.paperrecycles.org/about/paper-recycling-a-true-environmental-success-story; Scott Breen and Jay Siegel, "How a Unique Industry Collaboration Is Bottling a New Future for U.S. Glass Recycling," *GreenBiz*, October 12, 2018, https://www.greenbiz.com/article/how-unique-industry-collaboration-bottling-new-future-us-glass-recycling; Rick LeBlanc, "An Introduction to Metal Recycling," *The Balance Small Business*, updated June 25, 2019, https://www.thebalancesmb.com/an-introduction-to-metal-recycling-4057469.

6. Christopher Joyce, "U.S. Recycling Industry Is Struggling to Figure Out a Future without China," *All Things Considered*, National Public Radio, August 20, 2019, https://www.npr.org/2019/08/20/750864036/u-s-recycling-industry-is-struggling-to-figure-out-a-future-without-china.

7. "State by State Status of Legislation," B Lab, accessed October 28, 2020, https://benefitcorp.net/policymakers/state-by-state-status.

8. "B Corp Directory," B Lab, accessed October 28, 2020, https://bcorporation.net/directory/. This reference provides access to all B Corps in the section.

9. Rebecca Hamilton, "Staying True to Our Roots by Making Change," Badger Blog, W. S. Badger Company, September 22, 2016, https://blog.badgerbalm.com/staying-true-roots-making-change/.

10. Alana Semuels, "'Rampant Consumerism Is Not Attractive.' Patagonia Is Climbing to the Top—and Reimagining Capitalism Along the Way," *Time*, September 23, 2019, https://time.com/5684011/patagonia/.

11. "How Many Certified B Corps Are There Around the World?" B Lab, accessed October 28, 2020, https://bcorporation.net/faq-item/how-many-certified-b-corps-are-there-around-world.

12. Mikhail Davis, "Twenty Years Later Interface Looks Back at Ray Anderson's Legacy," *GreenBiz*, September 3, 2014, https://www.greenbiz.com/blog/2014/09/03/20-years-later-interface-looks-back-ray-andersons-legacy; "The Interface Story," Interface, accessed May 30, 2020, https://www.interface.com/US/en-US/sustainability/our-history-en_US.

13. "About the Institute," Cradle to Cradle Products Innovation Institute, accessed March 3, 2016, https://www.c2ccertified.org/about/about.

14. "Get Cradle to Cradle Certified," Cradle to Cradle Products Innovation Institute, accessed March 3, 2016, http://www.c2ccertified.org/get-certified/product-certification.

15. "Loll Designs Achieves Cradle to Cradle Certified™ Silver for Outdoor Furniture," MBDC, February 5, 2019, https://mbdc.com/loll-designs-achieves-cradle-cradle-certified-silver-outdoor-furniture/.

16. "Green Action," Playworld, accessed April 4, 2016, http://playworld.com/why_playworld/green_action.

17. "EIG and MBDC Assess the First Cradle to Cradle Certified Platinum Fabric for C&A," MBDC, February 7, 2020, https://mbdc.com/eig-and-mbdc-assess-the-first-cradle-to-cradle-certified-platinum-fabric-for-ca/.

18. "BioFoam," BEWi Synbra, accessed June 23, 2020, https://bewisynbra.com/product/biofoam/.

19. "How to Get Your Product Cradle to Cradle Certified™," MBDC, accessed April 9, 2020, https://mbdc.com/how-to-get-your-product-cradle-to-cradle-certified/.

20. Jeff Nesbit, "Oligarchy Nation," *U.S. News and World Report*, April 21, 2014, https://www.usnews.com/news/blogs/at-the-edge/2014/04/21/oligarchy-nation.

21. Jan Edwards, "Timeline of Personhood Rights and Powers," Women's International League for Peace and Freedom, revised June 2002, https://ratical.org/corporations/ToPRaP.html.

22. Jim Hightower and Phillip Frazier, eds., "How a Clerical Error Made Corporations 'People,'" *Hightower Lowdown*, April 2003, https://hightowerlowdown.org/node/664. Refers to reporting in Thom Hartmann, *Unequal Protection* (Emmaus, PA: Rodale, 2002).

23. Edwards, "Personhood Rights."

24. "Cosponsors: H.J.Res.48—116th Congress (2019–2020)," Library of Congress, accessed October 8, 2019, https://www.congress.gov/bill/116th-congress/house-joint-resolution/48/cosponsors?q=%7B%22search%22%3A%5B%22hjr48%22%5D%7D&s=2&r=1&overview=closed#tabs.

25. Karma Ura, Sabina Alkire, Tshoki Zangmo, and Karma Wangdi, *A Short Guide to Gross National Happiness Index* (Dechhog Lam, Thimphu, Bhutan: Center for Bhutan Studies, 2012), http://www.grossnationalhappiness.com/wp-content/uploads/2012/04/Short-GNH-Index-edited.pdf.

26. "Bhutan," *The Economist Intelligence Unit*, accessed January 16, 2020, http://country.eiu.com/bhutan.

27. Timothy Rybeck, "The UN Happiness Project," *New York Times*, March 28, 2012, https://www.nytimes.com/2012/03/29/opinion/the-un-happiness-project.html?auth=link-dismiss-google1tap.

28. John Helliwell, Richard Layard, and Jeffrey Sachs, eds. *World Happiness Report 2019* (New York: United Nations Sustainable Development Network, 2019), chapter 1, https://worldhappiness.report/ed/2019/#read.

29. Laura Musikanski et al., *2011 Happiness Report Card for Seattle* (publisher not identified), accessed January 16, 2020, http://wikiprogress.org/data/dataset/5514c3fe-3dd9-4631-984d-c6546a147b98/resource/54067a9c-329e-46c0-b65e-e70360effdbc/download/Seattle-HappinessReportCard-2011.pdf.

30. "Zero Waste Box Programs," TerraCycle, accessed June 24, 2020, https://www.terracycle.com/en-US/zero_waste_boxes.

31. "Cradle to Cradle Products Registry," Cradle to Cradle Products Innovation Institute, accessed April 9, 2020, https://www.c2ccertified.org/products/registry/search&material_for_product_designers=yes.

CHAPTER 19

1. Andrew Restuccia, "Greens: Climate March Breaks Record," *Politico*, September 21, 2014, https://www.politico.com/story/2014/09/peoples-climate-march-nyc-111177.

2. "What Is the United Nations Framework Convention on Climate Change," United Nations Climate Change, accessed October 29, 2020, https://unfccc.int/process-and-meetings/the-convention/what-is-the-united-nations-framework-convention-on-climate-change.

3. "Kyoto Protocol," United Nations Framework Convention on Climate Change, accessed May 1, 2015, http://unfccc.int/kyoto_protocol/items/2830.php. Michael Le Page, "Was Kyoto Climate Deal a Success? Figures Reveal Mixed Results," *New Scientist*, June 14, 2016, https://www.newscientist.com/article/2093579-was-kyoto-climate-deal-a-success-figures-reveal-mixed-results.

4. Al Gore, *Our Choice: A Plan to Solve the Climate Crisis* (Emmaus, PA: Rodale, 2009), 398–99.

5. "Climate Technology Center and Network," Climate Technology Center and Network, accessed October 29, 2017, https://www.ctc-n.org.

6. "Warsaw Outcomes," United Nations Framework Convention on Climate Change, accessed May 1, 2015, http://unfccc.int/key_steps/warsaw_outcomes/items/8006.php.

7. "History," Intergovernmental Panel for Climate Change, accessed May 1, 2015, http://ipcc.ch/organization/organization_history.shtml.

8. Martin Lukacs, "Indigenous Activists Take to Seine River to Protest Axing of Rights from Paris Climate Pact," *Guardian*, December 7, 2015, http://www.the

guardian.com/environment/true-north/2015/dec/07/indigenous-activists-take-to-seine-river-to-protest-axing-of-rights-from-paris-climate-pact.

9. Tyler Hamilton, "What Next After the Historic Paris Climate Change Agreement," *Toronto Star*, December 12, 2015, http://www.thestar.com/news/canada/2015/12/14/whats-next-after-the-historic-paris-climate-change-agreement.html.

10. Suzanne Goldenberg et al., "Paris Climate Deal: Nearly 200 Nations Sign an End of Fossil Fuel Era," *Guardian*, December 12, 2015, http://www.theguardian.com/environment/2015/dec/12/paris-climate-deal-200-nations-sign-finish-fossil-fuel-era.

11. Taylor Mayhall, "Does Mother Nature Get a Vote?" *University of Minnesota Law Review*, October 16, 2016, http://www.minnesotalawreview.org/2016/10/does-mother-nature-get-a-vote; Suzanne Goldenberg, "How US Negotiators Enabled Landmark Paris Climate Deal Was Republican Proof," *Guardian*, December 13, 2015, http://www.theguardian.com/us-news/2015/dec/13/climate-change-paris-deal-cop21-obama-administration-congress-republicans-environment.

12. "What Is Geo-engineering," Oxford Geoengineering Program, accessed November 23, 2019, http://www.geoengineering.ox.ac.uk/www.geoengineering.ox.ac.uk/what-is-geoengineering/what-is-geoengineering.

13. Fiona Harvey, "Christina Figueres: The Woman Tasked with Saving the World from Global Warming," *Guardian*, November 27, 2015, http://www.theguardian.com/environment/2015/nov/27/christiana-figueres-the-woman-tasked-with-saving-the-world-from-global-warming.

14. Karl Mathiesen and Fiona Harvey, "Climate Coalition Breaks Cover in Paris to Push for Binding and Ambitious Deal," *Guardian*, December 8, 2015, http://www.theguardian.com/environment/2015/dec/08/coalition-paris-push-for-binding-ambitious-climate-change-deal.

15. Arthur Neslen, "India Unveils Global Solar Alliance of 120 Countries at Paris Climate Summit," *Guardian*, November 30, 2015, http://www.theguardian.com/environment/2015/nov/30/india-set-to-unveil-global-solar-alliance-of-120-countries-at-paris-climate-summit.

16. Oliver Milman, "James Hansen, Father of Climate Change Awareness, Calls Paris Talks a 'Fraud,'" *Guardian*, December 12, 2015, http://www.theguardian.com/environment/2015/dec/12/james-hansen-climate-change-paris-talks-fraud.

17. "Agreed Climate Deal Offers a Frayed Lifeline for the World's Poorest People," Oxfam International, December 12, 2015, https://www.oxfam.org/en/press-releases/agreed-climate-deal-offers-frayed-life-line-worlds-poorest-people.

18. Megan Darby, "COP 21: NGO's React to the UN Paris Climate Deal," *Climate Home*, December 12, 2015, http://www.climatechangenews.com/2015/12/12/cop21-ngos-react-to-prospective-un-paris-climate-deal.

19. "350.org and Bill McKibben React to COP21 Climate Text," 350.org, December 12, 2020, https://350.org/press-release/cop21-reaction.

20. Pamela Falk, "Climate Negotiators Strike a Deal to Slow Global Warming," *CBS News*, updated December 12, 2015, https://www.cbsnews.com/news/cop21-climate-change-conference-final-draft-historic-plan.

21. Darby, "COP 21."

22. Fiona Harvey, "World Bank President Celebrates 'Game Changer' Paris Talks," *Guardian*, December 13, 2015, http://www.theguardian.com/business/2015/dec/13/world-bank-president-celebrates-game-changer-paris-talks.

23. "Outcomes of the Climate Change Conference in Marrakech," (fact sheet), Center for Climate and Energy Solutions, November 2016, https://www.c2es.org/international/neg"Cotiations/cop22-marrakech/summary.

24. Matt Payton, "Nearly 50 Countries Vow to Use 100% Renewable Energy by 2050," *Independent*, November 18, 2016, https://www.independent.co.uk/news/world/renewable-energy-target-climate-united-nations-climate-change-vulnerable-nations-ethiopia-a7425411.html.

25. Milman, "James Hansen."

26. Ker Than, "Estimated Social Cost of Climate Change Not Accurate, Stanford Scientists Say," *Stanford News*, August 4, 2015, http://news.stanford.edu/2015/01/12/emissions-social-costs-011215/.

27. Kristin Eberhard, "All the World's Carbon Pricing Systems in One Animated Map," *Sightline Daily*, November 17, 2014, http://daily.sightline.org/2014/11/17/all-the-worlds-carbon-pricing-systems-in-one-animated-map.

28. Carolyn Beeler, "Brittan Built an Empire Out of Coal, Now It's Giving It Up. Why Can't the U.S.?" *The World*, Public Radio International, June 18, 2018, https://www.pri.org/stories/2018-06-18/england-built-empire-out-coal-now-its-giving-it-why-can-t-us.

29. "The EU Emission Trading System (EU ETS)," European Commission, accessed January 17, 2020, http://ec.europa.eu/clima/policies/ets/index_en.htm.

30. "U.S. State Carbon Pricing Policies," Center for Climate and Energy Solutions, accessed October 10, 2019, https://www.c2es.org/document/us-state-carbon-pricing-policies.

31. "California Cap and Trade," Center for Climate and Energy Solutions, accessed October 10, 2019, https://www.c2es.org/content/california-cap-and-trade.

32. Robert Barnes, "Supreme Court: EPA Can Regulate Greenhouse Gasses, with Some Limits," *Washington Post*, June 23, 2014, https://www.washingtonpost.com/politics/supreme-court-limits-epas-ability-to-regulate-greenhouse-gas-emissions/2014/06/23/c56fc194-f1b1-11e3-914c-1fbd0614e2d4_story.html.

33. *State and Trends of Carbon Pricing 2020* (Washington, DC: World Bank, May 2020), p. 7, https://openknowledge.worldbank.org/bitstream/handle/10986/33809/9781464815867.pdf.

34. "Global Corporate Use of Carbon Pricing," CDP Worldwide, September 2014, https://www.cdp.net/CDPResults/global-price-on-carbon-report-2014.pdf.

35. "Santa Barbara Becomes 30th U.S. City to Commit to 100% Renewable Energy," (press release), Sierra Club, June 7, 2017, http://content.sierraclub.org/press-releases/2017/06/santa-barbara-becomes-30th-us-city-commit-100-renewable-energy.

36. "Cities and Climate Change," United Nations Environment Programme (UNEP), accessed October 29, 2020, https://www.unenvironment.org/explore-topics/resource-efficiency/what-we-do/cities/cities-and-climate-change.

37. Matt Richel, "San Diego Vows to Move Entirely to Renewable Energy in 20 Years," *New York Times*, December 15, 2015, https://www.nytimes.com/2015/12/16/science/san-diego-vows-to-move-entirely-to-renewable-energy-in-20-years.html.

38. "Green Cincinnati Plan," Office of Environment and Sustainability, City of Cincinnati, accessed August 25, 2018, https://www.cincinnati-oh.gov/oes/citywide-efforts/climate-protection-green-cincinnati-plan.

39. "Implementing Climate Action: Global Covenant of Mayors 2018 Global Aggregation Report," Global Covenant of Mayors, accessed October 11, 2019, https://www.globalcovenantofmayors.org/wp-content/uploads/2018/09/2018_GCOM_report_web.pdf.

40. "Sunnyside Landfill, Houston United States," C40 Cities, accessed October 11, 2019, https://www.c40reinventingcities.org/en/sites/holmes-road-land-fill-1271.html.

41. "East Garfield Park," C40 Cities, accessed October 11, 2019, https://www.c40reinventingcities.org/en/sites/east-garfield-park-1281.html.

42. "BLOXHUB Annual Report," BLOXHUB, April 29, 2019, https://cdn-bloxhub20.pressidium.com/wp-content/uploads/2020/08/BLOXHUB-Annual-Report-2018.pdf; "Urban Tech: Better Technologies for the Cities of the Future," BLOXHUB, August 31, 2020, https://bloxhub.org/news/urbantech-better-technology-for-the-cities-of-the-future.

43. Taryn Luna, "Plan to Power California with All Renewable Energy Heads to Jerry Brown," *Sacramento Bee*, August 28, 2018, https://www.sacbee.com/news/politics-government/capitol-alert/article217397360.html.

44. "Five Cities and States That Have Committed to Clean Energy," *Smart Grid*, accessed October 12, 2019, http://www.whatissmartgrid.org/featured-article/five-cities-and-states-that-have-committed-to-clean-energy.

45. Betsy Lilliam, "Cuomo Signs Historic Renewable Energy Bill into Law," *Solar Industry*, July 19, 2019, https://solarindustrymag.com/cuomo-signs-historic-renewable-energy-bill-into-law.

46. "Companies," *RE100*, accessed January 17, 2020, http://there100.org/companies.

47. "America's Top Colleges for Renewable Energy," Environment America Research and Policy Center, April 4, 2019, https://environmentamerica.org/reports/ame/americas-top-colleges-renewable-energy.

48. Koben Calhoun and Paul Bodnar, "Fulfilling America's Pledge," Rocky Mountain Institute, September 13, 2018, https://www.rmi.org/fulfilling-americas-pledge.

49. "Energy Innovation and Carbon Dividend Act," Citizen's Climate Lobby, accessed January 17, 2020, https://citizensclimatelobby.org/energy-innovation-and-carbon-dividend-act.

50. "Action: Download the 'Truth in 10' Slideshow," Climate Reality Project, accessed January 17, 2020, https://www.climaterealityproject.org/truth.

51. "Take Action," Union of Concerned Scientists, accessed January 17, 2020, https://www.ucsusa.org/take-action.

CHAPTER 20

1. Barbara Bradley Hagerty, "Quit Your Job: A Midlife Career Shift Can Be Good for Cognition, Well Being and Even Longevity," *The Atlantic*, April 2016, https://www.theatlantic.com/magazine/archive/2016/04/quit-your-job/471501/.

2. "Meet the Team," 350PDX, accessed August 10, 2019, https://350pdx.org/about-us/leadership-and-staff/.

3. Michael Brown, *Laying Waste: The Poisoning of America by Toxic Chemicals* (New York: Washington Square Press, 1981). Wikipedia, s.v. "Love Canal," accessed December 19, 2014, http://en.wikipedia.org/wiki/Love_Canal.

4. "What Is Superfund," U.S. Environmental Protection Agency, accessed January 21, 2020, https://www.epa.gov/superfund/what-superfund.

5. Bob Leonard, "Will COVID-19 Help Us Deal with Our Climate Crisis?" *Finite Earth Economy*, March 18, 2020, https://finiteeartheconomy.com/will-covid-19-help-us-deal-with-our-climate-crisis/.

CHAPTER 21

1. Milt Markewitz and Ruth Miller, *Language of Life* (Gleneden Beach, OR: Portal Center Press, 2013), 131–35.

2. "What Is the Paris Agreement?" United Nations Climate Change, accessed August 16, 2019, https://unfccc.int/process-and-meetings/the-paris-agreement/what-is-the-paris-agreement.

3. Wikipedia, s.v. "Akwesasne," accessed October 19, 2017, https://en.wikipedia.org/wiki/Akwesasne.

APPENDIX 1

1. "Summary of Solutions by Overall Rank," *Project Drawdown*, accessed November 23, 2019, https://www.drawdown.org/solutions-summary-by-rank.

SELECTED BIBLIOGRAPHY

Selected items reflect compiled information that can be accessed for in-depth study. For brevity, news reports, most blog stories, and most organizational websites are not included.

As You Sow. "Clean 200™." Accessed February 4, 2020. https://www.asyousow. org/clean200.

Audubon Society of Portland (Oregon). "Certification Criteria." Accessed August 8, 2014. http://audubonportland.org/issues/backyardhabitat/criteria.

B Lab. "B Corp Directory." Accessed October 30, 2020. https://bcorporation.net/ directory. https://benefitcorp.net/policymakers/state-by-state-status.

B Lab. "State by State Status of Legislation." Accessed October 30, 2020.

Bell, Ruth Greenspan, and Libor Jansky. "Public Participation in the Management of the Danube River: Necessary but Neglected." In *Public Participation in the Governance of Freshwater Resources*, edited by Carl Brush et al., 101–17. Tokyo: United Nations University Press, 2005.

Biemer, Jon, Willow Dixon, and Natalia Blackburn. "Our Environmental Handprint: The Good We Do." Institute of Electrical and Electronic Engineers, 2013 IEEE Conference on Technologies for Sustainability, Portland, Oregon.

Bonneville Power Administration. "BPA Distributed Energy Resource Benchmarking Report." December 2017. https://www.bpa.gov/EE/Technology/ demand-response/Documents/2017_Distributed_Energy_Resources_Bench marking_Report.pdf.

Brinkley, Douglas. *Rightful Heritage: Franklin D. Roosevelt and the Land of America.* New York: HarperCollins, 2016.

Brinkley, Douglas. *The Quiet World: Saving Alaska's Wilderness Kingdome 1879– 1960.* New York: HarperCollins, 2011.

Brodowsky, Pamela K. *Ecotourists Save the World.* New York: Penguin, 2010.

Brown, Michael. *Laying Waste: The Poisoning of America by Toxic Chemicals.* New York: Washington Square Press, 1981.

Buckler, Carolee, and Heather Creech. *Shaping the Future We Want: UN Decade for Education for Sustainable Development (2005–2014), Final Report.* United Nations Scientific, Educational, and Cultural Organization, 2014. http://unesdoc. unesco.org/images/0023/002301/230171e.pdf.

California Academy of Sciences and *National Geographic.* "iNaturalist." Accessed October 1, 2019. https://www.inaturalist.org.

Center for Agroforestry, University of Missouri. "What Is Agroforestry." Accessed July 13, 2017. http://www.centerforagroforestry.org/practices.

Central Asia Institute. "Why Your Support Matters." Accessed October 30, 2020. https://centralasiainstitute.org/why-it-matters.

City Repair Project. *Placemaking Guidebook,* 2nd ed. Portland, OR: City Repair Project, 2006.

Community Environmental Legal Defense Fund. *Community Rights Do-It-Yourself Guide to Lawmaking.* Accessed April 18, 2020. http://celdf.org/wp-content/uploads/2019/11/DIY-Guide-2019-FINAL-2.pdf.

Consumer Reports. "Electric Cars 101: The Answers to All Your EV Questions." July 21, 2019. https://www.consumerreports.org/hybrids-evs/electric-cars-101-the-answers-to-all-your-ev-questions.

Convention on International Trade in Endangered Species of Wild Fauna and Flora. *Convention on International Trade in Endangered Species of Wild Fauna and Flora* (text of treaty), signed 1973, amended 1979 and 1983. http://www.cites. org/eng/disc/text.php.

Cornell University. "Citizen Science." Accessed January 2, 2016. http://www. birds.cornell.edu/page.aspx?pid=1664.

Cradle to Cradle Products Innovation Institute. "Cradle to Cradle Products Registry." Accessed October 30, 2020. https://www.c2ccertified.org/products/registry.

D&R International. *2011 Building Energy Data Book.* Richland, Washington: Pacific Northwest National Laboratory, March 2012. https://ieer.org/wp/wp-content/uploads/2012/03/DOE-2011-Buildings-Energy-DataBook-BEDB.pdf.

Davis, Benjamin, Tony Dutzik, and Phineas Baxandall, *Transportation and the New Generation.* Frontier Group, U.S. PIRG. April 2012. https://uspirg.org/reports/usp/transportation-and-new-generation.

Dietrich, William. *The Final Forest: The Battle for the Last Great Trees of the Pacific Northwest.* New York: Penguin, 1992.

Earth Charter International. The Earth Charter Text. Accessed October 30, 2020. https://earthcharter.org/library/the-earth-charter-text.

Eat Low Carbon. "Diet Tips." Accessed August 24, 2018. http://eatlowcarbon. org/diet-tips.

Electronic Code of Federal Regulations. "National Organic Program. Title 7, Subtitle B, Chapter 1, Subchapter M, Part 205." April 22, 2020. https://www.ecfr. gov/cgi-bin/text-idx?tpl=/ecfrbrowse/Title07/7cfr205_main_02.tpl.

Endangered Species Coalition. "15 Ways to Help Protect Endangered Species." Accessed March 11, 2020. https://www.endangered.org/15-ways-to-help-protect-endangered-species.

Ensia. "Ensia Mentor Program." Accessed February 2, 2017. https://ensia.com/about/mentor-program.

Environmental Protection Agency. "Facts and Figures about Materials, Waste and Recycling." Accessed October 28, 2020. https://www.epa.gov/facts-and-figures-about-materials-waste-and-recycling/national-overview-facts-and-figures-materials#NationalPicture.

Federal Highway Administration, U.S. Department of Transportation. *Summary of Travel Trends: 2017 National Household Travel Survey*. July 2018. https://nhts.ornl.gov/assets/2017_nhts_summary_travel_trends.pdf.

Feeding America. "Feeding America, 2018 Annual Report." Accessed October 30, 2019. https://www.feedingamerica.org/about-us/financials.

Fink, Larry. "A Fundamental Reshaping of Finance" (letter to CEOs). BlackRock. Accessed January 23, 2020. https://www.blackrock.com/us/individual/larry-fink-ceo-letter.

Flow. *The Role of Walking and Cycling in Reducing Congestion, a Portfolio of Measures.* Brussels: FLOW Project, July 2016. http://h2020-flow.eu/uploads/tx_news/FLOW_REPORT_-_Portfolio_of_Measures_v_06_web.pdf.

Food and Agriculture Organization of the United Nations. *Greenhouse Gas Emissions from Ruminant Supply Chains.* Rome: United Nations, 2013. http://www.fao.org/docrep/018/i3461e/i3461e00.htm.

Food Not Lawns. "Directory of Local Chapters." Accessed June 20, 2016. http://www.foodnotlawns.com/local-chapters.html.

Garfield, Leanna. "10 Cities That Are Starting to Go Car-Free." *Business Insider*, August 16, 2016. https://www.businessinsider.com/cities-going-car-free-2016-8.

Gifford, Dawn. "20 Perennial Plants to Plant for Years of Bounty." Small Footprint Family. Accessed September 3, 2017. https://www.smallfootprintfamily.com/20-perennial-vegetables.

Gillio-Whitaker, Dina. *As Long as the Grass Grows.* Boston: Beacon Press, 2019.

Goddard, Sarah. "How to Pass a Plastic Bag Ban: 8 Key Lessons." *Green That Life.* March 17, 2019. https://greenthatlife.com/plastic-bag-ban.

Gore, Al. *Our Choice: A Plan to Solve the Climate Crisis.* Emmaus, PA: Rodale, 2009.

Hamerschlag, Kari. "Meat Eaters Guide: Report." Environmental Working Group, 2011. Accessed October 27, 2017. https://www.ewg.org/meateatersguide/a-meat-eaters-guide-to-climate-change-health-what-you-eat-matters/climate-and-environmental-impacts.

Hartmann, Thom. *Unequal Protection.* Emmaus, PA: Rodale, 2002.

Hawken, Paul. "Drawdown: The Most Comprehensive Plan Ever Proposed to Reverse Global Warming." YouTube video, 1:17, KODX Seattle. April 21, 2017. https://www.youtube.com/watch?v=KYvKv0lM-_A.

Helliwell, John, Richard Layard, and Jeffrey Sachs, eds. *World Happiness Report 2019*. New York: United Nations Sustainable Development Network, 2019. https://worldhappiness.report/ed/2019/#read.

Holmgren, David. *Permaculture, Principles and Pathways Beyond Sustainability*. Hepburn, Australia: Holmgren Design Services, 2002.

Juniper, Tony. *What Has Nature Ever Done for Us? How Money Really Does Grow on Trees* (Santa Fe, NM: Energetic Press, 2013).

Katz, Sandor, ed. *The Art of Fermentation*. White River Junction, VT: Chelsea Green, 2012.

Kiprop, Joseph. "Which Countries Have Banned Plastic Bags." *WorldAtlas*. Accessed August 29, 2018. https://www.worldatlas.com/articles/which-countries-have-banned-plastic-bags.html.

Klein, Naomi. *This Changes Everything*. New York: Simon & Schuster, 2014.

Kober, Amy Souers. "Twenty Years of Dam Removal Successes—and What's Up Next." American Rivers. June 27, 2019. https://www.americanrivers.org/2019/06/twenty-years-of-dam-removal-successes-and-whats-up-next.

Kraft, Michael, and Diana Wuertz. "Environmental Advocacy in the Corridors of Government." In *The Symbiotic Earth: Discourse and Our Creation of the Environment*, edited by James G. Cantrill and Christine L. Oravei. Lexington: University Press of Kentucky, 1996.

Langston, Bryce. "Living Big in a Tiny House." YouTube, *Living Big in a Tiny House*. Accessed January 21, 2020. https://www.youtube.com/user/livingbigtinyhouse/videos?app=desktop.

League of American Bicyclists. "Bicycling and Walking in the United States: 2018 Benchmarking Report." February 5, 2019. https://bikeleague.org/content/bicycling-and-walking-united-states-2018-benchmarking-report.

LeBlanc, Rick. "An Introduction to Metal Recycling." *The Balance Small Business*. Updated June 25, 2019. https://www.thebalancesmb.com/an-introduction-to-metal-recycling-4057469.

Leopold, Aldo. *A Sand County Almanac, with Essays on Conservation from Round River*. New York: Ballantine, 1974.

Local Harvest. "Community Supported Agriculture." Accessed November 22, 2019. https://www.localharvest.org/csa.

Louv, Richard. *Last Child in the Woods*. Chapel Hill, NC: Algonquin Books, 2008.

Lovins, Amory. "Hypercars: Uncompromised Vehicles, Disruptive Technologies, and the Rapid Transition to Hydrogen" (slide presentation). Presented to the California Air Resources Board, April 18, 2001. https://rmi.org/insight/hypercars-uncompromised-vehicles-disruptive-technologies-and-the-rapid-transition-to-hydrogen.

Lovins, Amory. *Reinventing Fire, Bold Business Solutions for the New Energy Era*. White River Junction, VT: Chelsea Green, 2011.

Lovins, Amory. *Small Is Profitable: The Hidden Economic Benefits of Making Electrical Resources the Right Size*. Basalt, CO: Rocky Mountain Institute, 2002.

MarineBio Conservation Society. "Marine Biology Degree Programs in the U.S. (by State)." Updated January 25, 2019. https://marinebio.org/careers/us-schools.

Markewitz, Milt, and Ruth Miller, *Language of Life*. Gleneden Beach, OR: Portal Center Press, 2013.

McKeown, Rosalyn. *Education for Sustainable Development Toolkit, Version 2.0.* Self-pub., July 2002. http://www.esdtoolkit.org.

McKibben, Bill. "Global Warming's Terrifying New Math." *Rolling Stone.* July 19, 2012. http://www.rollingstone.com/politics/news/global-warmings-terrifying-new-math-20120719.

Melosi, Martin. *Garbage in the Cities: Refuse, Reform, and the Environment.* Pittsburgh, PA: University of Pittsburgh Press, 2005.

Musikanski, Laura, et al. *2011 Happiness Report Card for Seattle.* Accessed January 16, 2020. http://wikiprogress.org/data/dataset/5514c3fe-3dd9-4631-984d-c6546a147b98/resource/54067a9c-329e-46c0-b65e-e70360effdbc/download/Seattle-HappinessReportCard-2011.pdf.

NARTS: The Professional Association of Resale Professionals. "Resale Industry and Statistics." Accessed July 30, 2019. http://www.narts.org/i4a/pages/index.cfm?pageid=3285.

National Agricultural Statistics Service, U.S. Department of Agriculture. "2014 and 2015 Organic Certifier Data." Accessed April 12, 2017. https://www.nass.usda.gov/Surveys/Guide_to_NASS_Surveys/Organic_Production/Organic_Certifiers/2016/USDA_Accredited_Certifying_Agent_Certified_Organic_Data_2014_2015.pdf.

National Conference of State Legislatures. "State Plastic and Paper Bag Legislation." Accessed January 13, 2020. https://www.ncsl.org/research/environment-and-natural-resources/plastic-bag-legislation.aspx.

National Oceanic and Atmospheric Administration. "National Marine Sanctuary System." Accessed August 24, 2014. http://sanctuaries.noaa.gov/.

National Oceanic and Atmospheric Administration. "Citizen Science at NOAA." Accessed January 13, 2019. https://oceanservice.noaa.gov/news/apr15/volunteer.html.

Native American College Fund. Accessed May 16, 2020. https://collegefund.org/about-us.

National Renewable Energy Laboratory. "Best Research Cell Efficiencies." Accessed October 25, 2020. https://www.nrel.gov/pv/cell-efficiency.html.

NOAA Fisheries, National Oceanic and Atmospheric Administration. "Marine Mammal Protection Act." Accessed October 30, 2020. https://www.fisheries.noaa.gov/topic/laws-policies#marine-mammal-protection-act.

Norris, Greg. "Introducing Handprints: A Net-Positive Approach to Sustainability." Harvard Extension School. Accessed December 21, 2019. https://www.extension.harvard.edu/introducing-handprints.

North American Association for Environmental Education. *K–12 Environmental Education: Guidelines for Excellence.* 2019. https://naaee.org/eepro/publication/excellence-environmental-education-guidelines-learning-k-12.

Northwest Energy Efficiency Alliance. "Areas of Work." Accessed April 27, 2020. https://neea.org/our-work.

Nye, Bill. *Unstoppable, Harnessing Science to Change the World.* New York: St. Martin's Press, 2015.

Ohlson, Kristin. *The Soil Will Save Us: How Scientists, Farmers, and Foodies Are Healing the Soil to Save the Planet.* New York: Rodale, 2014.

On Digital Marketing. "The 5 Customer Segments of Technology Adoption." Accessed August 22, 2019. https://ondigitalmarketing.com/learn/odm/foundations/5-customer-segments-technology-adoption.

Paper Recycles. "The Paper Recycling Success Story." Accessed April 11, 2020. https://www.paperrecycles.org/about/paper-recycling-a-true-environmental-success-story.

Perez, Caroline Criado. *Invisible Women: Data Bias in a World Designed for Men.* New York: Abrams, 2019.

Permaculture Worldwide Network. "Worldwide Permaculture Projects." Accessed November 22, 2019. https://permacultureglobal.org/projects.

Pipeline 101. "Where Are Liquids Pipelines Located?" Accessed September 28, 2017. http://www.pipeline101.org/Where-Are-Pipelines-Located.

Portland Tribune. "Portland Catches the Orange Line." September 2015.

Pradhan, Elina. "Female Education and Childbearing." Investing in Health, World Bank. Accessed February 9, 2017. http://blogs.worldbank.org/health/female-education-and-childbearing-closer-look-data.

Project Drawdown. "Summary of Solutions by Overall Rank." Accessed November 23, 2019. https://www.drawdown.org/solutions-summary-by-rank.

Railsback, Bruce, ed. *Creation Stories from Around the World*, 4th ed. Athens: University of Georgia, Department of Geology, July 2000. http://railsback.org/CS/CSFourCreations.html.

Regenerative Organic Alliance. "Framework for Regenerative Organic Certification." May 2018. https://regenorganic.org/wp-content/uploads/2018/06/ROC-Framework-1-May-2018.pdf.

Reijnders, Lucas, and Sam Soret. "Quantification of Environmental Impacts of Different Dietary Food Choices." *American Journal of Clinical Nutrition*, 2003. http://ajcn.nutrition.org/content/78/3/664S.full.

Rohwedder, Rocky. *Ecological Handprints.* Self-pub., 2016. https://ecologicalhandprints.atavist.com/ecological-handprints?promo.

Savory, Allan. "How to Fight Desertification and Reverse Climate Change." TED video, 22:13, February 2013. https://www.ted.com/talks/allan_savory_how_to_green_the_world_s_deserts_and_reverse_climate_change.

Schwartz, Judith. *Cows Save the Planet: And Other Improbable Ways of Restoring Soil to Heal the Earth.* White River Junction, VT: Chelsea Green, 2013.

Sea Grant, University of Hawaii. "Public Access Rights." Accessed December 9, 2016. http://seagrant.soest.hawaii.edu/public-access-rights.

Sharot, Tali. *The Influential Mind.* New York: Henry Holt, 2017.

Small Home Society. "Designers and Builders." Accessed September 9, 2015. http://smallhousesociety.net/designers-builders.

Staments, Paul. "Environmental Remediation Using Mycelium." YouTube video, 2:23, June 10, 2013. https://www.youtube.com/watch?v=lJHXkfNCl5E&noredirect=1.

Steffen, Alex, ed. *WorldChanging: A User's Guide for the 21st Century.* New York: Abrams, 2011.

Strickland, James. "Energy Efficiency of Different Modes of Transportation" (lecture notes). Stanford University. Updated February 12, 2009. http://adl.stanford.edu/aa260/Lecture_Notes_files/transport_fuel_consumption.pdf.

Surfrider Foundation. "Marine Debris Reauthorization Act." Accessed August 22, 2014. http://www.surfrider.org/campaigns/entry/marine-debris-reauthorization-act.

TerraCycle. "Zero Waste Box Programs." Accessed January 13, 2020. https://www.terracycle.com/en-US/zero_waste_boxes.Toensmeier, Eric. *Perennial Vegetables.* White River Junction, VT: Chelsea Green, 2007.

Toensmeier, Eric. *The Carbon Farming Solution.* White River Junction, VT: Chelsea Green, 2016.

United Nations Environment Programme. "Register of International Treaties and Other Agreements in the Field of the Environment." December 30, 2005. http://www.unep.org/delc/Portals/119/publications/register_Int_treaties_contents.pdf.

United Nations Framework Convention on Climate Change. "Background on the UNFCCC: The International Response to Climate Change." Accessed October 30, 2020. https://www.unenvironment.org/resources/report/register-international-treaties-and-other-agreements-field-environment.

United Nations Framework Convention on Climate Change. "Kyoto Protocol." Accessed May 1, 2015. http://unfccc.int/kyoto_protocol/items/2830.php.

United Nations Framework on Climate Change. "Adoption of the Paris Agreement." December 12, 2015. http://unfccc.int/resource/docs/2015/cop21/eng/l09.pdf.

United Nations General Assembly. *Transforming Our World: The 2030 Agenda for Sustainable Development.* September 25, 2015. http://www.un.org/ga/search/view_doc.asp?symbol=A/RES/70/1&Lang=E.

United Nations University, Institute for the Advanced Study of Sustainability. "Global RCI Network." Accessed February 19, 2020. http://www.rcenetwork.org/portal/rces-worldwide.

United States Geological Survey. "Bioremediation: Nature's Way to a Cleaner Environment." Accessed July 31, 2014. http://water.usgs.gov/wid/html/bioremed.html.

U.S. Department of Agriculture. "Local Food Directory: Farmers Market Directory." Updated December 11, 2019. https://www.ams.usda.gov/local-food-directories/farmersmarkets.

U.S. Energy Information Administration. "2015 Residential Energy Consumption Survey: Energy Consumption and Expenditures Tables." May 2018. https://www.eia.gov/consumption/residential/data/2015/c&e/pdf/ce1.1.pdf.

U.S. Environmental Protection Agency. "CERCLA Overview." Accessed August 21, 2014. http://www.epa.gov/superfund/policy/cercla.htm.

U.S. Environmental Protection Agency. "National Oil and Hazardous Substances Contingency Plan (NCP)." Accessed August 9, 2015. http://www2.epa.gov/emergency-response/national-oil-and-hazardous-substances-pollution-contingency-plan-ncp-overview.

U.S. Environmental Protection Agency. "National Overview: Facts and Figures on Materials, Wastes and Recycling." Accessed July 4, 2019. https://www.epa.gov/facts-and-figures-about-materials-waste-and-recycling/national-overview-facts-and-figures-materials.

U.S. Environmental Protection Agency. "National Priorities List (NPL)." Accessed October 30, 2020. https://www.epa.gov/superfund/superfund-national-priorities-list-npl.

U.S. Fish and Wildlife Service. "Lacey Act." Accessed March 11, 2020. https://www.fws.gov/international/laws-treaties-agreements/us-conservation-laws/lacey-act.html.

Ura, Karma, Sabina Alkire, Tshoki Zangmo, and Karma Wangdi. *A Short Guide to Gross National Happiness Index.* Dechhog Lam, Bhutan: Center for Bhutan Studies, 2012. http://www.grossnationalhappiness.com/wp-content/uploads/2012/04/Short-GNH-Index-edited.pdf.

Wackernagel, Mathis, and Bert Beyers. *Ecological Footprint: Managing the Biocapacity Budget.* Gabriola Island, BC: New Society. 2019.

Wallace, Aubrey. *Eco-Heroes, Twelve Tales of Environmental Victory.* San Francisco, CA: Mercury House, 1993.

Williams, Julie Steward. *From the Mountains to the Sea, Early Hawaiian Life.* Honolulu, HI: Kamehameha Schools Press, 1997. http://ulukau.org/elib/cgi-bin/library?c=english&l=en.

World Bank. *State and Trends of Carbon Pricing 2019.* Washington, DC: World Bank, 2019. https://openknowledge.worldbank.org/handle/10986/31755.

World Wind Energy Association. "Windpower Capacity Worldwide Reaches 597 GW, 50.1 Added in 2018." February 25, 2019. https://wwindea.org/blog/2019/02/25/wind-power-capacity-worldwide-reaches-600-gw-539-gw-added-in-2018.

World Wind Energy Association. "World Market Has Reached 486 Gigawatts from Where 54 GW Has Been Installed Last Year." Statistics. June 8, 2017. http://www.wwindea.org/11961-2.

INDEX

ABOUT THE AUTHOR

Jon R. Biemer is a mechanical engineer and holds a certificate in process-oriented psychology. For twenty-three years, he conducted research and managed programs for the Energy Efficiency group of Bonneville Power Administration. He now provides organizational development consulting to environmental nonprofits.

Jon and his wife, Willow, have adopted many of the lifestyle practices highlighted in *Our Environmental Handprints*. Their eco-remodeled home was featured on the Tour of Green Homes for Portland, Oregon. They turned their lawn into a food forest. They have lived without a car for thirteen years.

Jon's forty-year career in sustainability includes facilitating the nation's first industrial energy-efficiency conference, walking the proposed Keystone XL pipeline route from Alberta to Nebraska, and organizing local Pachamama Alliance programs.